Rare Diseases of the Immune System

Marina II
Laura Maddii Emmi (private collection)

Alessandro Plebani • Vassilios Lougaris
Editors

Agammaglobulinemia

Springer

Editors
Alessandro Plebani
Pediatrics Clinic
Department of Clinical and Experimental Sciences
University of Brescia and Spedali Civili of Brescia
Brescia
Italy

Vassilios Lougaris
Pediatrics Clinic
Department of Clinical and Experimental Sciences
University of Brescia and Spedali Civili of Brescia
Brescia
Italy

Series editors
Lorenzo Emmi
Department of Experimental and Clinical Medicine
University of Florence
Florence
Italy

Medical Pathology Unit
Center for Autoimmune Systemic Diseases
A.O.U. Careggi Hospital
Florence
Italy

Domenico Prisco
Department of Experimental and Clinical Medicine
University of Florence
Florence
Italy

Medical Pathology Unit
Center for Autoimmune Systemic Diseases
A.O.U. Careggi Hospital
Florence
Italy

ISSN 2282-6505 ISSN 2283-6403 (electronic)
Rare Diseases of the Immune System
ISBN 978-3-319-22713-9 ISBN 978-3-319-22714-6 (eBook)
DOI 10.1007/978-3-319-22714-6

Library of Congress Control Number: 2015953197

Springer Cham Heidelberg New York Dordrecht London

Printed on acid-free paper

Springer International Publishing AG Switzerland is part of Springer Science+Business Media (www.springer.com)

To Antonella, Marta, Irene, Valeria,
Annamaria, Emanuele

Preface

Agammaglobulinemia was first described in 1952 by Colonel Odgen C. Bruton on a single male pediatric patient with recurrent invasive infections sustained by encapsulated bacteria by utilizing the serum protein electrophoresis that had just been developed at the Walter Reed Army Hospital in Bethesda, Maryland, USA. Additional immunological assays (immunoelettrophoresis, Ouchterlony double immunodiffusion, radial immunodiffusion techniques) allowing for the assessment of different immunoglobulin isotypes led to the identification of immunoglobulin deficiency disorders of one or more of these isotypes, named dysgammaglobulinemias, with agammaglobulinemia being part of this group. At that time, nothing was known on T and B lymphocytes, which were only described in the mid-1960s. This innovative discovery led to a progressive explosion of interest in the field of immunology and together with the advances in the field of cellular (lymphocyte culture, production of monoclonal antibodies) and molecular biology (linkage analysis, candidate gene analysis, next-generation sequencing) allowed to better define the pathogenetic mechanisms of several humoral immunofeficiencies, among them agammaglobulinemia. Agammaglobulinemia, in its different forms, is the object of this monograph, in which clinical and therapeutic issues integrate with principles of basic and molecular immunology, as an effort to offer an updated overview of this disorder. We hope that this monograph may offer the notions of basic immunology to clinicians and useful clinical information, to researchers in order for both to integrate their knowledge in this field with the final aim to improve the clinical assistance to patients affected with agammaglobulinemia.

We would like to thank all expert senior and junior scientists and clinicians from around the world that contributed to this book with their expertise, their time, and efforts, without which this monograph would not have been accomplished. We would also like to thank Springer for making this book possible.

Brescia, Italy

Alessandro Plebani
Vassilios Lougaris

Contents

Early B Cell Development

1

Hermann Eibel

1.1 From Hematopoietic Stem Cells to B Cell Progenitors

The homeostasis, maintenance, and differentiation of hematopoietic stem cells (HSCs) take place within the highly specialized microenvironment of the human bone marrow. These processes are regulated on the one side by interactions with stroma cells of non-hematopoietic origin [64, 71] and on the other side by cell type-specific intrinsic programs controlled by the timed expression of transcription factors combined with epigenetic changes of the chromatin structure [10, 23, 119]. Since HSCs are multipotent, they can differentiate into cells of the lymphocyte, monocyte/myeloid, and erythrocyte lineage. Moreover, expressing cytokine receptors such as Flt3 and c-kit, pattern-recognition receptors like TLRs, Rig I-like receptors (RLRs), Nod-like receptors (NLRs), and purinergic receptors [72, 93], HSCs integrate and respond to various signals generated by their microenvironment. Therefore, stress, tissue damage, inflammation, and infections may change the composition of the HSC pool [22, 86, 89].

Repopulation studies carried out in mice show that HSCs are separated into subsets with long-, intermediate-, and short-term repopulation capacity ([52] Benveniste, 2010 #1385). The distribution of HSCs into these subsets is regulated by multiple interactions with osteoblasts, the vasculature, and stroma cells [18]. In humans, HSCs are mainly characterized according to the expression of the surface markers CD34, CD133, CD90, and CD49f (integrin α6) and the lack of CD45RA and CD38 expression [26]. Expression of the latter markers and

H. Eibel
Center of Chronic Immunodeficiency, University Medical Center Freiburg, Engesser str. 4, Freiburg 79108, Germany
e-mail: hermann.eibel@uniklinik-freiburg.de

A. Plebani, V. Lougaris (eds.), *Agammaglobulinemia*,
Rare Diseases of the Immune System 4, DOI 10.1007/978-3-319-22714-6_1

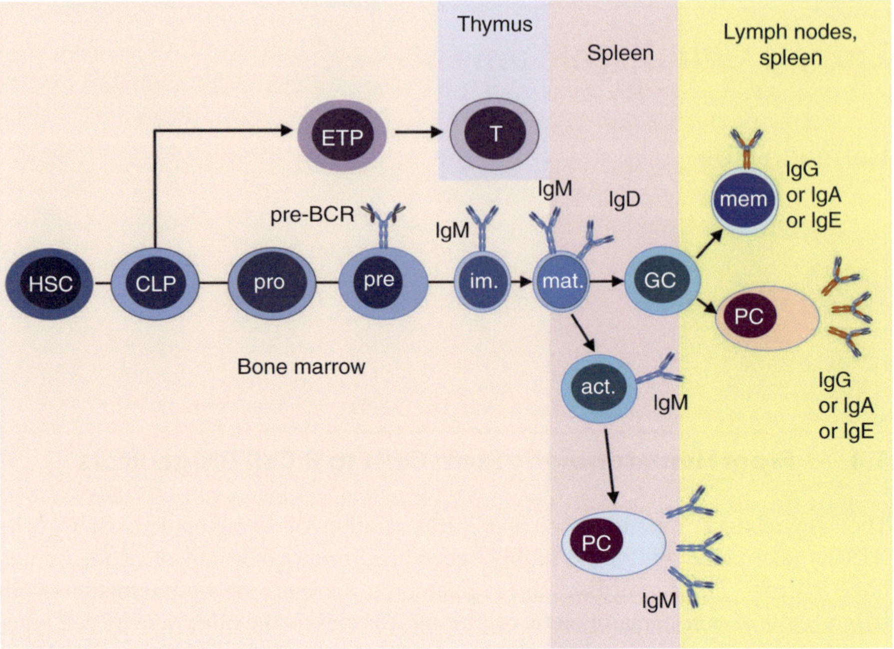

Fig. 1.1 An overview on B cell development. B cell precursors (pro-B, pre-B, and immature B cells) develop in the bone marrow from common lymphoid progenitors (*CLP*) that originate from hematopoietic stem cells (*HSC*). CLP may also emigrate as early thymocyte progenitor cells from the bone marrow and home to the thymus where they develop into mature T cells. Immature IgM+ B cells exit from the bone marrow and home as IgM+ IgD+ transitional B cells to the spleen and eventually also to gut-associated lymphoid tissues (*GALT*). There they complete maturation as naive IgM+ IgD+ follicular or marginal zone B cells (*mat.*). Antigen binding activates B cells which may in a context-dependent manner either become, together with T helper cells, founders of germinal centers where they undergo somatic hypermutation of variable regions and class-switch recombination of their constant parts. In the course of the germinal center reaction, B cells responding to antigen develop into IgG$^+$ or IgA$^+$ switched memory B cells (*mem*) and into long-lived plasma cells (*PC*) secreting high amounts of IgG and IgA. If B cells are activated by T-independent antigens, they may develop outside of the follicles into IgM-secreting short-lived plasma cells

downregulation of CD90 and CD49f are generally used to identify more differentiated subsets such as common lymphoid progenitors (CLP). CD34$^+$ CD45$^+$CD10$^+$ CD7$^{+/-}$ expressing CLPs initiate the differentiation of lymphoid lineage and develop into B, T, and NK cells. However, they may also differentiate into monocytes although with reduced efficiency [26, 52].

In contrast to the early thymic progenitors (ETPs) that are leaving the bone marrow to the thymus where they differentiate into mature T cells, B cell progenitors stay in the bone marrow where they develop into IgM-expressing immature B cells (Fig. 1.1).

1.2 A Complex Network of Transcription Factors Regulates the Commitment to the B Cell Lineage and Pro-B and Pre-B Cell Development

The different phases of B lymphocyte development are governed by the regulated expression and activity of proteins that are essential signaling components of cytokine and antigen receptor signaling pathways as well as by enzymes that catalyze the rearrangement of the immunoglobulin (Ig) VDJ and VJ gene segments.

Combined with the expression of specific cell surface and cytoplasmic markers, V(D)J recombination is used to distinguish distinct subpopulations of developing B cell (Fig. 1.2) [9, 54]. The somatic recombination events of the rearrangement process that joins the V, D, and J segments of the H chain and the VJ segments of the immunoglobulin L chain are catalyzed by two recombinases encoded by the

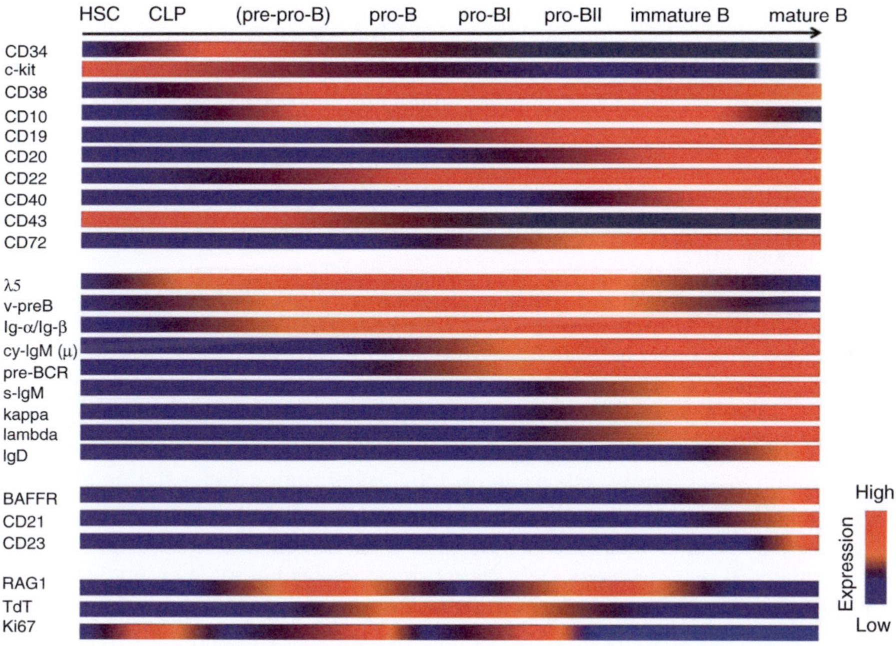

Fig. 1.2 Phases and markers of early B cell development. The time line shows the expression levels of stage-specific surface markers and receptors (CD34, c-kit, CD38, CD10, CD19, CD20, CD22, CD40, CD43, CD72), of components of the pre-BCR and the BCR (λ5, Vpre-B, Ig-α, Ig-β, the cytoplasmic μ-chain, surface IgM, kappa and lambda light chains, and IgD) of markers and receptors expressed by transitional and mature B cells (BAFFR, CD21, CD23), of enzymes involved in immunoglobulin gene rearrangement (RAG1, TdT), and of the proliferation marker Ki67. Expression levels are indicated to the right and are compiled from Bendall et al [9] and Kraus et al. [9, 54]

recombination activating genes *RAG1* and *RAG2*. *RAG* genes are expressed first in $CD34^+$ early pro-B cells (also called pre-pro-B cells), and their expression coincides with the phosphorylation of STAT5 in response to IL-7 binding to the IL-7R [9]. In mouse CLPs, IL-7R expression seems to be regulated by the transcription factor Miz-1 as B cell precursors of mice lacking the POZ (poxvirus and zinc finger) domain of Miz-1 cannot respond to IL-7. Miz-1 has a dual function by acting as repressor of the JAK inhibitor "suppressor of cytokine signaling 1" (Socs1) and as transactivator of Bcl-2 expression. Since Miz-1-deficient B cell precursors fail to induce Bcl-2 expression, these cells are more susceptible to die by apoptosis. As a consequence of impaired IL-7R signaling, *Miz1−/−* pro-B cells express very low levels of two transcription factors that are essential inducers of B cell development: E2A (Tcf3) and early B cell factor 1 (EBF1). Therefore, in Miz-1-deficient mice, B cell development is arrested at the pro-B cell stage [53].

Induced by E2A activity, upregulation of EBF1 expression is one of the most important steps in early B cell development. The central role of EBF1 as an opener for B cell development becomes readily apparent if EBF1 expression is ablated in knockout mice as in these animals the loss of EBF1 function arrests B cell development before pro-B cells start to rearrange the H-chain gene segments [58]. These data suggested that EBF1 is a central component of the concerted action of transcription factors that control the expression of *Rag* genes in mice. In fact, in pro- and pre-B cells, EBF1 binds to the regulatory sequences of more than 500 genes. Upon EBF1 binding, chromatin changes allow the transcriptional activation of target genes or poising of target loci for later expression at subsequent stages of B cell differentiation [107].

The expression of EBF1 is regulated by two other transcription factors expressed in HSC and in common lymphocyte progenitors, namely, E2A [77] and RUNX1 [98]. E2A does not only activate transcription of the *Ebf1* gene but it also cooperates with this transcription factor by synergistically upregulating the expression of the recombinases RAG1 and RAG2, of the pre-BCR and BCR-associated signal transducer protein Ig-α (CD79A), and of PAX5 [77]. PAX5, a member of the paired box transcription factor family, acts as an essential transcription factor during the commitment to the B lineage as well as in all later stages of B cell development [14, 75].

RUNX1 (also known as AML-1) plays an essential role in normal hematopoietic development. It recognizes its DNA target sequences as homodimer, but binding is enhanced more than tenfold if it forms a heterodimer with the core binding factor β (CBFβ) [101]. These RUNX1-CBF complexes regulate expression of *EBF1* by epigenetic mechanisms. In *Runx1*-deficient pro-B cells, repressive histones surrounding the *Ebf1* promoter carry heavy H3K27 methylation, whereas binding of RUNX1-CBF complexes to the *Ebf1* proximal promoter supports opening of its chromatin structure and induces EBF1 expression [98].

In addition to E2A, RUNX1, Miz-1, and EBF1, factors of the IKAROS family [70] play a crucial role in early B cell progenitors. Together with histone deacetylases (HDAC) and the DNA-dependent ATPase, Mi-2 IKAROS forms the so-called Mi-2/NuRD (nucleosome remodeling deacetylase) complex. Binding of the Mi-2/NuRD complex to target genes needs a specific guide that recognizes the target

DNA. By recruiting Mi-2/HDAC to specific regions of heterochromatin at or close to regulatory sequences in genes, IKAROS proteins are restructuring the chromatin at specific loci during lymphocyte differentiation, thus poising lymphocyte-specific genes—similar to EBF1—for later transcription [50]. If IKAROS is lacking, the NuRD complex activates other genes that are transcriptionally poised by common transcription factors and encode many proteins regulating cellular metabolism and proliferation. As these genes are not targets of IKAROS, mutations or deletions of the IKAROS-encoding *Izkf1* gene prevent lymphocyte maturation and promote the progression to a transformed state and leukemogenesis. Thus, IKAROS plays an important role in preventing the transformation of B cell precursors by engaging functionally opposing epigenetic and genetic networks [96]. In the course of normal B cell development, IKAROS regulates the expression of lambda5 (CD179B, λ5) together with EBF1 [105] and is required to induce the expression of the *RAG1* and *RAG2 genes* [91]. Moreover, IKAROS and the closely related transcription factor AIOLOS bind to the c-Myc promoter and directly inhibit c-Myc expression that is induced by IL-7 binding to the IL-7 receptor. As c-Myc repression is required for the upregulation of the cell cycle kinase inhibitor $p27^{KIP}$ and for the repression of the cyclin-dependent kinase regulator cyclin D3, the IKAROS transcription factors function as key elements in regulating the expansion of the pro- and pre-B cell pool [60]. Similar to the other transcription factors playing a key role in early B cell development, IKAROS1, encoded by the *IZKF1* gene, is not expressed constitutively but is regulated by another transcription factor called FOXO1. FOXO1 is a member of the forkhead transcription factor family. These factors are involved in many cellular processes including lymphomagenesis, tumor suppression, and cell death, as well as the long-term maintenance of stem cells [4]. In B cell precursors, FOXO1 controls the splicing of *Izkf1* mRNA [2].

Like EBF1 and IKAROS, the ETS transcription factor PU.1 contributes to the commitment to the B cell lineage [16]. PU.1 upregulates together with the transcription factor SpiB expression of Bruton's tyrosine kinase (BTK), which in turn acts as negative regulator of IL-7R-dependent STAT5 activation. As IL-7R signaling and STAT5 activation are needed for the development of mouse B cells, low PU.1 expression levels promote lymphocyte development, whereas high levels of PU.1 foster the development into myeloid cells [19, 61].

Controlled by these transcription factors, PAX5 acts as master regulator of B cell development and of maintaining essential B cell functions [75]. PAX5 commits precursor cells to the B cell lineage by activating the expression of B cell-specific genes and by repressing non-B lymphoid genes [75, 92]. PAX5 is expressed almost exclusively in B cells with the exception of cells in the testis and in the developing CNS [1]. In the B cell lineage, PAX5 expression starts at the pro-B cell stage and lasts until B cells develop into plasma cells [20]. Induced by EBF1 [87, 107], PAX5 activity drives the progression of B cell development beyond the pro-B cell stage. For example, PAX5 controls in a positive feedback loop together with EBF1 and E2A/E47 the expression of *CD19, SLP65 (BLNK), and CD79A* [13]. PAX5-deficient pro-B cells remain uncommitted to the B cell lineage, and exposure to the appropriate microenvironment allows the development of *Pax5–/–* pro-B cells into functional T cells, macrophages, osteoclasts, dendritic cells, granulocytes, and natural killer cells [65, 75, 92]. In mature B cells, PAX5 activity is needed to keep the cells on the B cell lineage track as conditional inactivation of *Pax5* in mature B cells changes the gene expression pattern allowing the expression on non-B lymphoid genes [41]. Interestingly, the loss of B cell-specific gene expression (e.g., of *Bcl6*) is paralleled by upregulation of transcription factors expressed normally in plasma cells such as Blimp-1 (*Prdm1*) and of Xbp1 as well as the secretion of IgM [74]. Thus, PAX5 expression prevents the transition of B cells into the plasma cell stage unless activation of B cells in response to antigen leads to the expression of plasma cell-specific master genes like Blimp-1 at levels that are sufficient to repress *Pax5* transcription and to switch to the plasma cell program [47, 76].

In humans, immunodeficiencies resulting from mutations in *PU.1*, *RUNX1*, *MIZ1*, *EBF1*, *AIOLOS*, *PAX5*, or E2A (*TCF3*) were not found so far. A mutation in the DNA-binding Zn finger of IKAROS caused pancytopenia [36], whereas mutations in PAX5 [99] and fusions between E2A [56] or RUNX1 [42] and other genes have frequently been described in acute lymphoblastic pre-B cell leukemia.

1.3 Assembly and Signaling of the Pre-B Cell Receptor Complex

$CD10^+$ pre-pro-B cells start to express CD79A (Ig-α) and CD79B (Ig-β) that are both essential signal-transducing components of the pre-B cell and B cell antigen receptor complex [39, 90]. In addition, these cells express CD179A (Vpre-B) and CD179B (λ5) that serve combined as surrogate light chains (SLC) before the "real" light chains become available after the productive rearrangement of the immunoglobulin kappa and lambda L-chain gene segments. Before the onset of D-J H-recombination, the recombination activating genes encoding the RAG1 and RAG2 proteins are expressed together with the gene encoding the enzyme terminal deoxynucleotidyl transferase (TdT) [35, 111, 116]. While the recombinases RAG1 and RAG2 are indispensable for somatic VDJ gene segment rearrangement [79, 95], TdT adds nucleotides at the joints between rearranged gene segments in a template-independent manner to increase the variability and the repertoire of the rearranged μ-H chains. After completing VDJ rearrangement, human B cell progenitors have developed into $CD19^+$ $CD20^+$ pre-B I [111]. If these cells express a functional VDJ-Cμ immunoglobulin H chain, the μ-H chains form the pre-BCR complex together with the signal-transducing components Ig-α and Ig-β, the surrogate light chain, CD19, and BCAP ([80] Saito, 2003 #1307). CD179A (Vpre-B) and CD179B (lambda5 or λ5) associate as surrogate light chain that is bound in a non-covalent manner to the V region and the first constant domain of the μ-H chain [48, 49]. This signaling complex regulates the most critical steps in B cell development, namely, an initial shutdown of RAG expression and H-chain rearrangement (allelic exclusion), proliferative expansion of pre-B cells that express a functional pre-BCR, and the induction of L-chain rearrangement during a second phase of RAG expression.

At the intracellular side of the pre-BCR, the adapter protein SLP65 (BLNK) [118] plays an essential role in early B cell development as it forms an attachment site for several key enzymes that regulate B cell differentiation. In *Blnk*-deficient mice, pre-B cells accumulate and only a fraction of these cells develops into mature B cells. The latter show reduced Ca^{2+} mobilization and impaired responses to antigen [45]. Different from BLNK-deficient humans [55, 68] that have a severe block at the pre-B cell stage and lack almost all mature B cells, *Blnk*-deficient mice contain mature B cells and accumulate pre-B cells due to the steady expansion of their pre-B cell compartment. Moreover, *Blnk*-deficient mice frequently develop pre-B cell leukemia that can be prevented by reintroducing SLP65 expression in precursor B cells [31] . These striking differences between BLNK-deficient human and mouse pre-B cells are most likely attributed to the constitutively active JAK3/STAT5 signaling pathway in SLP65/BLNK-deficient mouse pre-B cells. This signaling pathway is induced by autocrine production of IL-7 binding to their IL-7R [73]. Different from mice, IL-7R-dependent activation of STAT5 in human B cell precursors occurs before the expression of the pre-BCR [9]. This uncouples the initial proliferation phase induced in CLP/pro-B cells by binding of IL-7 to IL-7R from the signaling phase induced in late pro-B and pre-BI cells upon assembly of a functional pre-BCR and minimizes the risk that mutations in single genes encoding components of the pre-BCR result in uncontrolled proliferation of pre-B cells and leukemogenesis.

Although it has been found that the pre-BCR binds to galectin [11, 27, 29, 34] or to heparin sulfates [12]—these molecules are present on the surface of mouse stroma cells and may serve as external ligands—it also has been clearly demonstrated that activation of pre-B cells via the pre-BCR is induced in a ligand-independent manner. Binding of the surrogate light chain to carbohydrates linked to a conserved asparagine residue at position 46 of the first constant domain of the μ-heavy chain allows homo-aggregation of the pre-BCR and autonomous signaling by concentrating the intracellular signaling components coupled to the pre-BCR [109]. As a result of pre-BCR aggregation, Src-related kinases can phosphorylate SYK, which in turn phosphorylates its substrate proteins including SLP65. This adapter protein is then recruited to the plasma membrane by its *N*-terminal leucine zipper motif [51]. Phosphorylated by SYK, the SLP65 adapter protein provides a highly specialized scaffold for several key enzymes linked to the signaling pathways downstream of pre-BCR and of the BCR. Therefore, B cell development is arrested at the pro-B/pre-B cell stage in *Syk*-deficient mice [94].

One of the signaling pathways coupled via SLP65 to the pre-BCR is the activation of class I phosphoinositide 3-kinases (PI3K). These enzymes catalyze the synthesis of phosphatidylinositol 3-phosphate, phosphatidylinositol (3,4)-bisphosphate (PIP_2), and phosphatidylinositol (3,4,5)-trisphosphate (PIP_3). To form a functional enzyme, the regulatory subunit (p85), which is encoded by the *PIK3R1* gene, has to associate with one of the three isoforms (α, β, or δ) of the catalytic subunit (p110). The p110α, p110β, and p110δ subunits are encoded by separate genes (*PIK3CA, PIK3CB, and PIK3CD*). Whereas the p110α and p110β isoforms are ubiquitously expressed, the p110δ is primarily found in lymphocytes [81] including pro-B and pre-B cells [8]. Deletion of the p110δ encoding *Pik3cd* gene in mice arrests B cell development at the stage of pro-B cells resulting in decreased numbers of mature B cells, disturbed Ca^{2+} flux, proliferation, impaired immune responses to T-independent, as well as T-dependent antigens and agammaglobulinemia [44]. Immature B cells from p110δ-deficient mice do not downregulate RAG expression and continue to rearrange L-chain loci demonstrating that PI3K signaling plays an important role in regulating allelic exclusion of immunoglobulin L chains [59]. As p110α partially compensates the lack of p110δ in mice, some of the B cell precursors can develop up to the mature B cell stage. If both isoforms are lacking, the developmental arrest is more pronounced while RAG expression is not shut down by pre-BCR signaling. The third isoform, p110β, does not seem to play a role in B cell development [88]. This is different from mutations in the *PIK3R1* gene encoding the PI3K structural subunit p85 that cause a partial block in B cell development with a phenotype that is similar to p110δ or *Btk*-deficient mice [32].

Phosphorylated PI3K activates AKT (protein kinase B) [81], which in turn leads to the phosphorylation and inactivation of the transcription factors FOXO1 and FOXO3A [38]. As FOXO1 is needed for the correct splicing of the IKAROS-encoding mRNA, homo-oligomerization of the pre-BCR and activation of the downstream signaling pathways directly regulate in a negative feedback loop the rearrangement of the heavy gene segments [2].

Another, equally important pathway linked via SLP65 to the pre-BCR involves activation of phospholipase Cγ2 (PLCγ2) through phosphorylation catalyzed by BTK [5, 37, 103]. PLCγ2 hydrolyzes phospholipids of the plasma membrane into diacylglycerol (DAG) and inositol 1,4,5-trisphosphate (IP3). In mice lacking the PLCγ2-encoding *Plcg2* gene, B cell development is blocked like in *Blnk*-deficient mice at the stage of pre-B cells. Although the double deletion of *Blnk* and of *Plcg2* has an additive effect by interrupting normal B cell development at the stage between pro- and pre-B cells, H-chain rearrangement continues in the double-knockout mice showing that allelic exclusion and VDJ rearrangement occur independently of both molecules and do not acquire pre-BCR signaling and aggregation [102, 120].

As pointed out above, expression of the pre-BCR and its autonomous signaling is the first crucial checkpoint in B cell maturation. In humans, mutations abolishing the function of any of the single components of the pre-BCR [25, 30, 66, 67] and of its downstream signaling molecules [21, 55, 68, 108, 113] or the rearrangement of immunoglobulin gene segments [97, 114] block B cell development at the stage of pro-B or pre-B cells. For example, humans carrying mutations in *BTK* or *SLP65* [68, 108] lack circulating B cells and antibodies of all isotypes.

1.4 Allelic Exclusion and Receptor Editing

As each single B cell should express only one type of antibody composed of a unique H and L chain, signals created by the assembly of the functionally rearranged VDJ μ-H chain in the pre-BCR complex downregulate RAG expression [33, 104]. Allelic exclusion preventing rearrangement of the second H-chain allele also limits the critical phase that permits chromosomal translocations caused by the DNA double-strand breaks introduced into the H-chain locus during the rearrangement process. The potentially dangerous activity of the RAG recombinases is regulated on posttranslational as well as on transcriptional levels. In pre-B cells and developing thymocytes, RAG2 accumulates during the G1 phase of the cell cycle but is degraded when cells enter the S and G2/M phase. RAG2 degradation depends on phosphorylation of the threonine residue 490 by cyclin-dependent kinases. Thus, DNA double-strand breaks and the intermediates of VDJ recombination accumulate during the G0/G1 phase and are resolved before the cells start to replicate their chromosomes [57]. In addition to the regulation by cell cycle-dependent RAG2 degradation, allelic exclusion is controlled directly by the dsDNA breaks that are introduced by RAG activity at the recombination signal sequences flanking the V,D, and J gene segments. Rearrangement of one allele recruits and activates the serine/threonine kinase "ataxia telangiectasia mutated" (ATM) leading to the positioning of the other allele to pericentromeric heterochromatin. This mechanism prevents biallelic recombination and recombination-induced chromosome breaks [40]. As a second safeguard, ATM activates the serine/threonine kinase PIM2. Its smaller isoform acts as cell cycle repressor and prevents pro-B and pre-B cells from entering the S phase [7]. Nevertheless, RAG-induced chromosomal aberrations can occur spontaneously leading to the development of acute lymphoblastic leukemia [82].

After having rearranged the H-chain alleles, pre-B cells undergo a few rounds of replication and start to re-express the RAG genes to allow rearrangement of the L-chain V and J gene segments. This transition from replicating to differentiating pre-B cells is regulated by interferences between pre-BCR and STAT5 signaling [78] and changes in the activity of PI3K and of its substrate proteins [6, 38, 78] combined with the upregulation of FOXO [3] and AIOLOS/IKAROS proteins [60]

Similar to allelic exclusion at the μHC locus, rearrangement of the light-chain loci κ and λ occurs at consecutive stages. Rearrangement is first induced only at one allele of one light-chain locus, which is in general the κ-locus. If the rearrangement is successful leading to a functional light chain that pairs well with the heavy chain, the pre-B cell proceeds to the next stage, the IgM$^+$ immature B cell. If the recombination of L-chain V and J gene segments fails or if the light chain does not pair well with the H chain, rearrangement starts at the second allele or at the Igλ locus [28]. If these rearrangements at the L-chain loci allow the expression of functional κ- or λ-L chains, they form together with the μ-H-chain IgM that is expressed on the cell surface of immature CD19$^+$ CD10$^+$ CD38$^+$ CD24$^+$ B cells.

At this stage, the B cell is ready to leave the bone marrow through the sinusoids and to enter the bloodstream that transports the cells into the spleen where they complete the early stage of development as mature, naïve B cells.

The exit from the bone marrow is controlled by two G protein-coupled chemokine receptors: cannabinoid receptor 2 (CB2) and sphingosine-1-phosphate receptor 1 (S1P1). Activation of CB2 by endocannabinoids and interactions between integrin $\alpha_4\beta_1$ (VLA-4) and VCAM-1 increase the retention time of immature B cells in the bone marrow [84], whereas S1P1 activation induced by high S1P concentrations in the bloodstream promotes the exit into circulation [85, 100].

If immature B cells express autoreactive B cell receptors or H and L chains that pair not efficiently enough, the increased retention time in the parenchyma and in the sinusoids of the bone marrow allows secondary rearrangements of the light-chain loci. The replacement of functional light chains by secondary rearrangements ("receptor editing") was discovered first in transgenic or knockin mouse models expressing rearranged self-reactive H- and L-chain genes [17, 83, 106]. Receptor editing is based on the continuous expression of RAG proteins that catalyze the somatic recombination between those V and J segments of the κ- or λ-loci that are located either upstream (V) or downstream of the previously rearranged VJ gene segments.

Binding to self-antigen to surface IgM expressed on immature B cells seems to activate RAP1 GTPases [43]. Their activity may enforce adhesion to parenchymal cells and retention in the bone marrow. Additionally, sustained RAG expression is promoted by canonical NF-κB activity that is upregulated by binding of (self-)antigens to the BCR [112]. Therefore, increased NF-κB activity and nuclear translocation of p50/relA heterodimers mark cells actively undergoing receptor editing [15]. The same process of secondary light-chain rearrangements is also found in humans [46, 62, 63].

Compared to binding of cognate antigens to surface IgM and IgD, the BCR signals induced by binding to self-antigens to surface IgM under physiological conditions seem to be weak, and only very low levels of BCR-dependent signaling seem to be required to induce secondary rearrangements in immature B cells in the bone marrow. This was demonstrated by marking immature B cells that were passing the short phase of transitional B cells while they were entering the spleen before reaching the stage of mature B cells with an inducible GFP gene controlled by Nur77 regulatory sequences [121]. Nur77 transcription is activated in B and T cells when these cells strongly bind to antigen [24, 69]. Therefore, it can be used to label B cells that were in contact with antigens cross-linking their BCR. Using these Nur77-GFP reporter mice, it became clear that almost all transitional B cells and mature B cells in the mouse spleen were contacting (self-)antigen with different binding intensities [121]. Comparing the signaling properties between IgM and IgD, it became clear that only IgM is capable of responding to low-valence antigens including monovalent antigens while activation through IgD needs higher cross-linking levels or presentation of antigens in the form of immune complexes [110]. Thus, many of the IgM^{low} IgD^{hi} mature follicular B cells marked by Nur77-GFP may have downregulated surface IgM after they have been in contact with low-valence antigens. As these cells still respond to antigen presented as immune complexes, they form a large repertoire of B cells that have the potential to develop in T-dependent immune responses into switched memory B cells and plasma cells [110].

Recent data provided evidence that immature B cells may have pass or may have even developed in the intestinal lamina propria before reaching the stage of mature B cells. In fact, the lamina propria of the small intestine in mice harbors RAG-expressing pro-B, pre-B, and immature B lineage cells. The immature B cells from the lamina propria expressed different proportions of immunoglobulin κ- and λ-L chains than the immature B cells developing in the bone marrow.

Therefore, not only the "sterile" environment of the bone marrow but also the gut microbiome with its signals from commensal bacteria shapes the repertoire of the B cell compartment [117]. In humans it has been shown that a major subset of immature B cells emigrates from the bone marrow and homes to gut-associated lymphoid tissues [115]. Most of these cells were found to be activated suggesting that they were in contact with antigens that were either binding to surface IgM or to toll-like receptors. Interestingly, immature B cells from systemic lupus erythematosus patients seemed not to be homing to the gut suggesting that gut homing and encounter with gut-derived antigens have a regulatory function that may protect against autoimmunity. It is not known so far which of the different primary antibody deficiencies is linked to defects in gut homing of immature B cells. But future work may show if changes in gut homing are associated with IgA deficiency and if the so far unexplained autoimmune manifestations found in primary antibody deficiencies relate to impaired selection of autoimmune B cells in the bone marrow or in gut-associated tissues.

References

1. Adams B, Dorfler P, Aguzzi A, Kozmik Z, Urbanek P, Maurer-Fogy I, Busslinger M (1992) Pax-5 encodes the transcription factor BSAP and is expressed in B lymphocytes, the developing CNS, and adult testis. Genes Dev 6:1589–1607
2. Alkhatib A, Werner M, Hug E, Herzog S, Eschbach C, Faraidun H, Kohler F, Wossning T, Jumaa H (2012) FoxO1 induces Ikaros splicing to promote immunoglobulin gene recombination. J Exp Med 209:395–406
3. Amin RH, Schlissel MS (2008) Foxo1 directly regulates the transcription of recombination-activating genes during B cell development. Nat Immunol 9:613–622
4. Arden KC (2007) FoxOs in tumor suppression and stem cell maintenance. Cell 128: 235–237
5. Baba Y, Hashimoto S, Matsushita M, Watanabe D, Kishimoto T, Kurosaki T, Tsukada S (2001) BLNK mediates Syk-dependent Btk activation. Proc Natl Acad Sci U S A 98: 2582–2586
6. Baracho GV, Cato MH, Zhu Z, Jaren OR, Hobeika E, Reth M, Rickert RC (2014) PDK1 regulates B cell differentiation and homeostasis. Proc Natl Acad Sci U S A 111:9573–9578
7. Bednarski JJ, Nickless A, Bhattacharya D, Amin RH, Schlissel MS, Sleckman BP (2012) RAG-induced DNA double-strand breaks signal through Pim2 to promote pre-B cell survival and limit proliferation. J Exp Med 209:11–17
8. Beer-Hammer S, Zebedin E, von Holleben M, Alferink J, Reis B, Dresing P, Degrandi D, Scheu S, Hirsch E, Sexl V et al (2010) The catalytic PI3K isoforms p110gamma and p110delta contribute to B cell development and maintenance, transformation, and proliferation. J Leukoc Biol 87:1083–1095
9. Bendall SC, Davis KL, el Amir AD, Tadmor MD, Simonds EF, Chen TJ, Shenfeld DK, Nolan GP, Pe'er D (2014) Single-cell trajectory detection uncovers progression and regulatory coordination in human B cell development. Cell 157:714–725
10. Boller S, Grosschedl R (2014) The regulatory network of B-cell differentiation: a focused view of early B-cell factor 1 function. Immunol Rev 261:102–115
11. Bonzi J, Bornet O, Betzi S, Kasper BT, Mahal LK, Mancini SJ, Schiff C, Sebban-Kreuzer C, Guerlesquin F, Elantak L (2015) Pre-B cell receptor binding to galectin-1 modifies galectin-1/carbohydrate affinity to modulate specific galectin-1/glycan lattice interactions. Nat Commun 6:6194

12. Bradl H, Wittmann J, Milius D, Vettermann C, Jack HM (2003) Interaction of murine precursor B cell receptor with stroma cells is controlled by the unique tail of lambda 5 and stroma cell-associated heparan sulfate. J Immunol 171:2338–2348
13. Busslinger M (2004) Transcriptional control of early B cell development. Annu Rev Immunol 22:55–79
14. Busslinger M, Nutt SL, Rolink AG (2000) Lineage commitment in lymphopoiesis. Curr Opin Immunol 12:151–158
15. Cadera EJ, Wan F, Amin RH, Nolla H, Lenardo MJ, Schlissel MS (2009) NF-kappaB activity marks cells engaged in receptor editing. J Exp Med 206:1803–1816
16. Carotta S, Wu L, Nutt SL (2010) Surprising new roles for PU.1 in the adaptive immune response. Immunol Rev 238:63–75
17. Chen C, Nagy Z, Prak EL, Weigert M (1995) Immunoglobulin heavy chain gene replacement: a mechanism of receptor editing. Immunity 3:747–755
18. Chitteti BR, Kobayashi M, Cheng Y, Zhang H, Poteat BA, Broxmeyer HE, Pelus LM, Hanenberg H, Zollman A, Kamocka MM et al (2014) CD166 regulates human and murine hematopoietic stem cells and the hematopoietic niche. Blood 124:519–529
19. Christie DA, Xu LS, Turkistany SA, Solomon LA, Li SK, Yim E, Welch I, Bell GI, Hess DA, DeKoter RP (2015) PU.1 opposes IL-7-dependent proliferation of developing B cells with involvement of the direct target gene bruton tyrosine kinase. J Immunol 194:595–605
20. Cobaleda C, Schebesta A, Delogu A, Busslinger M (2007) Pax5: the guardian of B cell identity and function. Nat Immunol 8:463–470
21. Conley ME, Dobbs AK, Quintana AM, Bosompem A, Wang YD, Coustan-Smith E, Smith AM, Perez EE, Murray PJ (2012) Agammaglobulinemia and absent B lineage cells in a patient lacking the p85alpha subunit of PI3K. J Exp Med 209:463–470
22. Coskun S, Chao H, Vasavada H, Heydari K, Gonzales N, Zhou X, de Crombrugghe B, Hirschi KK (2014) Development of the fetal bone marrow niche and regulation of HSC quiescence and homing ability by emerging osteolineage cells. Cell Rep 9:581–590
23. Dege C, Hagman J (2014) Mi-2/NuRD chromatin remodeling complexes regulate B and T-lymphocyte development and function. Immunol Rev 261:126–140
24. Dinkel A, Warnatz K, Ledermann B, Rolink A, Zipfel PF, Burki K, Eibel H (1998) The transcription factor early growth response 1 (Egr-1) advances differentiation of pre-B and immature B cells. J Exp Med 188:2215–2224
25. Dobbs AK, Yang T, Farmer D, Kager L, Parolini O, Conley ME (2007) Cutting edge: a hypomorphic mutation in Igbeta (CD79b) in a patient with immunodeficiency and a leaky defect in B cell development. J Immunol 179:2055–2059
26. Doulatov S, Notta F, Laurenti E, Dick JE (2012) Hematopoiesis: a human perspective. Cell Stem Cell 10:120–136
27. Elantak L, Espeli M, Boned A, Bornet O, Bonzi J, Gauthier L, Feracci M, Roche P, Guerlesquin F, Schiff C (2012) Structural basis for galectin-1-dependent pre-B cell receptor (pre-BCR) activation. J Biol Chem 287:44703–44713
28. Engel H, Rolink A, Weiss S (1999) B cells are programmed to activate kappa and lambda for rearrangement at consecutive developmental stages. Eur J Immunol 29:2167–2176
29. Espeli M, Mancini SJ, Breton C, Poirier F, Schiff C (2009) Impaired B-cell development at the pre-BII-cell stage in galectin-1-deficient mice due to inefficient pre-BII/stromal cell interactions. Blood 113:5878–5886
30. Ferrari S, Lougaris V, Caraffi S, Zuntini R, Yang J, Soresina A, Meini A, Cazzola G, Rossi C, Reth M, Plebani A (2007) Mutations of the Igb gene cause agammaglobulinemia in man. J Exp Med 204:2047–2051
31. Flemming A, Brummer T, Reth M, Jumaa H (2003) The adaptor protein SLP-65 acts as a tumor suppressor that limits pre-B cell expansion. Nat Immunol 4:38–43
32. Fruman DA, Snapper SB, Yballe CM, Davidson L, Yu JY, Alt FW, Cantley LC (1999) Impaired B cell development and proliferation in absence of phosphoinositide 3-kinase p85alpha. Science 283:393–397

33. Galler GR, Mundt C, Parker M, Pelanda R, Martensson IL, Winkler TH (2004) Surface mu heavy chain signals down-regulation of the V(D)J-recombinase machinery in the absence of surrogate light chain components. J Exp Med 199:1523–1532
34. Gauthier L, Rossi B, Roux F, Termine E, Schiff C (2002) Galectin-1 is a stromal cell ligand of the pre-B cell receptor (BCR) implicated in synapse formation between pre-B and stromal cells and in pre-BCR triggering. Proc Natl Acad Sci U S A 99:13014–13019
35. Ghia P, ten Boekel E, Sanz E, de la Hera A, Rolink A, Melchers F (1996) Ordering of human bone marrow B lymphocyte precursors by single-cell polymerase chain reaction analyses of the rearrangement status of the immunoglobulin H and L chain gene loci. J Exp Med 184:2217–2229
36. Goldman FD, Gurel Z, Al-Zubeidi D, Fried AJ, Icardi M, Song C, Dovat S (2012) Congenital pancytopenia and absence of B lymphocytes in a neonate with a mutation in the Ikaros gene. Pediatr Blood Cancer 58:591–597
37. Hashimoto S, Iwamatsu A, Ishiai M, Okawa K, Yamadori T, Matsushita M, Baba Y, Kishimoto T, Kurosaki T, Tsukada S (1999) Identification of the SH2 domain binding protein of Bruton's tyrosine kinase as BLNK--functional significance of Btk-SH2 domain in B-cell antigen receptor-coupled calcium signaling. Blood 94:2357–2364
38. Herzog S, Hug E, Meixlsperger S, Paik JH, DePinho RA, Reth M, Jumaa H (2008) SLP-65 regulates immunoglobulin light chain gene recombination through the PI(3)K-PKB-Foxo pathway. Nat Immunol 9:623–631
39. Herzog S, Reth M, Jumaa H (2009) Regulation of B-cell proliferation and differentiation by pre-B-cell receptor signalling. Nat Rev Immunol 9:195–205
40. Hewitt SL, Yin B, Ji Y, Chaumeil J, Marszalek K, Tenthorey J, Salvagiotto G, Steinel N, Ramsey LB, Ghysdael J et al (2009) RAG-1 and ATM coordinate monoallelic recombination and nuclear positioning of immunoglobulin loci. Nat Immunol 10:655–664
41. Horcher M, Souabni A, Busslinger M (2001) Pax5/BSAP maintains the identity of B cells in late B lymphopoiesis. Immunity 14:779–790
42. Hormaeche I, Licht JD (2007) Chromatin modulation by oncogenic transcription factors: new complexity, new therapeutic targets. Cancer Cell 11:475–478
43. Ishida D, Su L, Tamura A, Katayama Y, Kawai Y, Wang SF, Taniwaki M, Hamazaki Y, Hattori M, Minato N (2006) Rap1 signal controls B cell receptor repertoire and generation of self-reactive B1a cells. Immunity 24:417–427
44. Jou ST, Carpino N, Takahashi Y, Piekorz R, Chao JR, Carpino N, Wang D, Ihle JN (2002) Essential, nonredundant role for the phosphoinositide 3-kinase p110delta in signaling by the B-cell receptor complex. Mol Cell Biol 22:8580–8591
45. Jumaa H, Wollscheid B, Mitterer M, Wienands J, Reth M, Nielsen PJ (1999) Abnormal development and function of B lymphocytes in mice deficient for the signaling adaptor protein SLP-65. Immunity 11:547–554
46. Kalinina O, Wang Y, Sia K, Radic M, Cazenave PA, Weigert M (2014) Light chain editors of anti-DNA receptors in human B cells. J Exp Med 211:357–364
47. Kallies A, Hasbold J, Tarlinton DM, Dietrich W, Corcoran LM, Hodgkin PD, Nutt SL (2004) Plasma cell ontogeny defined by quantitative changes in blimp-1 expression. J Exp Med 200:967–977
48. Karasuyama H, Rolink A, Melchers F (1993) A complex of glycoproteins is associated with VpreB/lambda 5 surrogate light chain on the surface of mu heavy chain-negative early precursor B cell lines. J Exp Med 178:469–478
49. Karasuyama H, Rolink A, Shinkai Y, Young F, Alt FW, Melchers F (1994) The expression of Vpre-B/lambda 5 surrogate light chain in early bone marrow precursor B cells of normal and B cell-deficient mutant mice. Cell 77:133–143
50. Kim J, Sif S, Jones B, Jackson A, Koipally J, Heller E, Winandy S, Viel A, Sawyer A, Ikeda T et al (1999) Ikaros DNA-binding proteins direct formation of chromatin remodeling complexes in lymphocytes. Immunity 10:345–355
51. Kohler F, Storch B, Kulathu Y, Herzog S, Kuppig S, Reth M, Jumaa H (2005) A leucine zipper in the N terminus confers membrane association to SLP-65. Nat Immunol 6:204–210

52. Kondo M, Weissman IL, Akashi K (1997) Identification of clonogenic common lymphoid progenitors in mouse bone marrow. Cell 91:661–672
53. Kosan C, Saba I, Godmann M, Herold S, Herkert B, Eilers M, Moroy T (2010) Transcription factor miz-1 is required to regulate interleukin-7 receptor signaling at early commitment stages of B cell differentiation. Immunity 33:917–928
54. Kraus H, Kaiser S, Aumann K, Bonelt P, Salzer U, Vestweber D, Erlacher M, Kunze M, Burger M, Pieper K et al (2014) A feeder-free differentiation system identifies autonomously proliferating B cell precursors in human bone marrow. J Immunol 192:1044–1054
55. Lagresle-Peyrou C, Millili M, Luce S, Boned A, Sadek H, Rouiller J, Frange P, Cros G, Cavazzana M, Andre-Schmutz I et al (2014) The BLNK adaptor protein has a nonredundant role in human B-cell differentiation. J Allergy Clin Immunol 134:145–154
56. LeBrun DP (2003) E2A basic helix-loop-helix transcription factors in human leukemia. Front Biosci J Virtual Lib 8:s206–s222
57. Li Z, Dordai DI, Lee J, Desiderio S (1996) A conserved degradation signal regulates RAG-2 accumulation during cell division and links V(D)J recombination to the cell cycle. Immunity 5:575–589
58. Lin H, Grosschedl R (1995) Failure of B-cell differentiation in mice lacking the transcription factor EBF. Nature 376:263–267
59. Llorian M, Stamataki Z, Hill S, Turner M, Martensson IL (2007) The PI3K p110delta is required for down-regulation of RAG expression in immature B cells. J Immunol 178:1981–1985
60. Ma S, Pathak S, Mandal M, Trinh L, Clark MR, Lu R (2010) Ikaros and Aiolos inhibit pre-B-cell proliferation by directly suppressing c-Myc expression. Mol Cell Biol 30:4149–4158
61. Medina KL, Pongubala JM, Reddy KL, Lancki DW, Dekoter R, Kieslinger M, Grosschedl R, Singh H (2004) Assembling a gene regulatory network for specification of the B cell fate. Dev Cell 7:607–617
62. Meffre E, Davis E, Schiff C, Cunningham-Rundles C, Ivashkiv LB, Staudt LM, Young JW, Nussenzweig MC (2000) Circulating human B cells that express surrogate light chains and edited receptors. Nat Immunol 1:207–213
63. Meffre E, Schaefer A, Wardemann H, Wilson P, Davis E, Nussenzweig MC (2004) Surrogate light chain expressing human peripheral B cells produce self-reactive antibodies. J Exp Med 199:145–150
64. Mendelson A, Frenette PS (2014) Hematopoietic stem cell niche maintenance during homeostasis and regeneration. Nat Med 20:833–846
65. Mikkola I, Heavey B, Horcher M, Busslinger M (2002) Reversion of B cell commitment upon loss of Pax5 expression. Science 297:110–113
66. Minegishi Y, Coustan-Smith E, Rapalus L, Ersoy F, Campana D, Conley ME (1999) Mutations in Igalpha (CD79a) result in a complete block in B-cell development. J Clin Invest 104:1115–1121
67. Minegishi Y, Coustan-Smith E, Wang YH, Cooper MD, Campana D, Conley ME (1998) Mutations in the human lambda5/14.1 gene result in B cell deficiency and agammaglobulinemia. J Exp Med 187:71–77
68. Minegishi Y, Rohrer J, Coustan-Smith E, Lederman HM, Pappu R, Campana D, Chan AC, Conley ME (1999) An essential role for BLNK in human B cell development. Science 286:1954–1957
69. Mittelstadt PR, DeFranco AL (1993) Induction of early response genes by cross-linking membrane Ig on B lymphocytes. J Immunol 150:4822–4832
70. Molnar A, Georgopoulos K (1994) The Ikaros gene encodes a family of functionally diverse zinc finger DNA-binding proteins. Mol Cell Biol 14:8292–8303
71. Morrison SJ, Scadden DT (2014) The bone marrow niche for haematopoietic stem cells. Nature 505:327–334
72. Nagai Y, Garrett KP, Ohta S, Bahrun U, Kouro T, Akira S, Takatsu K, Kincade PW (2006) Toll-like receptors on hematopoietic progenitor cells stimulate innate immune system replenishment. Immunity 24:801–812

73. Nakayama J, Yamamoto M, Hayashi K, Satoh H, Bundo K, Kubo M, Goitsuka R, Farrar MA, Kitamura D (2009) BLNK suppresses pre-B-cell leukemogenesis through inhibition of JAK3. Blood 113:1483–1492
74. Nera KP, Kohonen P, Narvi E, Peippo A, Mustonen L, Terho P, Koskela K, Buerstedde JM, Lassila O (2006) Loss of Pax5 promotes plasma cell differentiation. Immunity 24:283–293
75. Nutt SL, Heavey B, Rolink AG, Busslinger M (1999) Commitment to the B-lymphoid lineage depends on the transcription factor Pax5. Nature 401:556–562
76. Nutt SL, Kallies A, Belz GT (2008) Blimp-1 connects the intrinsic and extrinsic regulation of T cell homeostasis. J Clin Immunol 28:97–106
77. O'Riordan M, Grosschedl R (1999) Coordinate regulation of B cell differentiation by the transcription factors EBF and E2A. Immunity 11:21–31
78. Ochiai K, Maienschein-Cline M, Mandal M, Triggs JR, Bertolino E, Sciammas R, Dinner AR, Clark MR, Singh H (2012) A self-reinforcing regulatory network triggered by limiting IL-7 activates pre-BCR signaling and differentiation. Nat Immunol 13:300–307
79. Oettinger MA, Schatz DG, Gorka C, Baltimore D (1990) RAG-1 and RAG-2, adjacent genes that synergistically activate V(D)J recombination. Science 248:1517–1523
80. Okada T, Maeda A, Iwamatsu A, Gotoh K, Kurosaki T (2000) BCAP: the tyrosine kinase substrate that connects B cell receptor to phosphoinositide 3-kinase activation. Immunity 13:817–827
81. Okkenhaug K (2013) Signaling by the phosphoinositide 3-kinase family in immune cells. Annu Rev Immunol 31:675–704
82. Papaemmanuil E, Rapado I, Li Y, Potter NE, Wedge DC, Tubio J, Alexandrov LB, Van Loo P, Cooke SL, Marshall J et al (2014) RAG-mediated recombination is the predominant driver of oncogenic rearrangement in ETV6-RUNX1 acute lymphoblastic leukemia. Nat Genet 46:116–125
83. Pelanda R, Schwers S, Sonoda E, Torres RM, Nemazee D, Rajewsky K (1997) Receptor editing in a transgenic mouse model: site, efficiency, and role in B cell tolerance and antibody diversification. Immunity 7:765–775
84. Pereira JP, An J, Xu Y, Huang Y, Cyster JG (2009) Cannabinoid receptor 2 mediates the retention of immature B cells in bone marrow sinusoids. Nat Immunol 10:403–411
85. Pereira JP, Xu Y, Cyster JG (2010) A role for S1P and S1P1 in immature-B cell egress from mouse bone marrow. PLoS One 5, e9277
86. Pietras EM, Lakshminarasimhan R, Techner JM, Fong S, Flach J, Binnewies M, Passegue E (2014) Re-entry into quiescence protects hematopoietic stem cells from the killing effect of chronic exposure to type I interferons. J Exp Med 211:245–262
87. Pongubala JM, Northrup DL, Lancki DW, Medina KL, Treiber T, Bertolino E, Thomas M, Grosschedl R, Allman D, Singh H (2008) Transcription factor EBF restricts alternative lineage options and promotes B cell fate commitment independently of Pax5. Nat Immunol 9:203–215
88. Ramadani F, Bolland DJ, Garcon F, Emery JL, Vanhaesebroeck B, Corcoran AE, Okkenhaug K (2010) The PI3K isoforms p110alpha and p110delta are essential for pre-B cell receptor signaling and B cell development. Sci Signal 3:ra60
89. Rashidi NM, Scott MK, Scherf N, Krinner A, Kalchschmidt JS, Gounaris K, Selkirk ME, Roeder I, Lo Celso C (2014) In vivo time-lapse imaging shows diverse niche engagement by quiescent and naturally activated hematopoietic stem cells. Blood 124:79–83
90. Reth M, Nielsen P (2014) Signaling circuits in early B-cell development. Adv Immunol 122:129–175
91. Reynaud D, Demarco IA, Reddy KL, Schjerven H, Bertolino E, Chen Z, Smale ST, Winandy S, Singh H (2008) Regulation of B cell fate commitment and immunoglobulin heavy-chain gene rearrangements by Ikaros. Nat Immunol 9:927–936
92. Rolink AG, Nutt SL, Melchers F, Busslinger M (1999) Long-term in vivo reconstitution of T-cell development by Pax5-deficient B-cell progenitors. Nature 401:603–606
93. Rossi L, Salvestrini V, Ferrari D, Di Virgilio F, Lemoli RM (2012) The sixth sense: hematopoietic stem cells detect danger through purinergic signaling. Blood 120:2365–2375

94. Saijo K, Schmedt C, Su IH, Karasuyama H, Lowell CA, Reth M, Adachi T, Patke A, Santana A, Tarakhovsky A (2003) Essential role of Src-family protein tyrosine kinases in NF-kappaB activation during B cell development. Nat Immunol 4:274–279
95. Schatz DG, Oettinger MA, Schlissel MS (1992) V(D)J recombination: molecular biology and regulation. Annu Rev Immunol 10:359–383
96. Schjerven H, McLaughlin J, Arenzana TL, Frietze S, Cheng D, Wadsworth SE, Lawson GW, Bensinger SJ, Farnham PJ, Witte ON et al (2013) Selective regulation of lymphopoiesis and leukemogenesis by individual zinc fingers of Ikaros. Nat Immunol 14:1073–1083
97. Schwarz K, Bartram CR (1996) V(D)J recombination pathology. Adv Immunol 61:285–326
98. Seo W, Ikawa T, Kawamoto H, Taniuchi I (2012) Runx1-Cbfbeta facilitates early B lymphocyte development by regulating expression of Ebf1. J Exp Med 209:1255–1262
99. Shah S, Schrader KA, Waanders E, Timms AE, Vijai J, Miething C, Wechsler J, Yang J, Hayes J, Klein RJ et al (2013) A recurrent germline PAX5 mutation confers susceptibility to pre-B cell acute lymphoblastic leukemia. Nat Genet 45:1226–1231
100. Sic H, Kraus H, Madl J, Flittner KA, von Munchow AL, Pieper K, Rizzi M, Kienzler AK, Ayata K, Rauer S et al (2014) Sphingosine-1-phosphate receptors control B-cell migration through signaling components associated with primary immunodeficiencies, chronic lymphocytic leukemia, and multiple sclerosis. J Allergy Clin Immunol 134:420–428
101. Speck NA (2001) Core binding factor and its role in normal hematopoietic development. Curr Opin Hematol 8:192–196
102. Su YW, Flemming A, Wossning T, Hobeika E, Reth M, Jumaa H (2003) Identification of a pre-BCR lacking surrogate light chain. J Exp Med 198:1699–1706
103. Tan JE, Wong SC, Gan SK, Xu S, Lam KP (2001) The adaptor protein BLNK is required for B cell antigen receptor-induced activation of nuclear factor-kappa B and cell cycle entry and survival of B lymphocytes. J Biol Chem 276:20055–20063
104. ten Boekel E, Melchers F, Rolink AG (1998) Precursor B cells showing H chain allelic inclusion display allelic exclusion at the level of pre-B cell receptor surface expression. Immunity 8:199–207
105. Thompson EC, Cobb BS, Sabbattini P, Meixlsperger S, Parelho V, Liberg D, Taylor B, Dillon N, Georgopoulos K, Jumaa H et al (2007) Ikaros DNA-binding proteins as integral components of B cell developmental-stage-specific regulatory circuits. Immunity 26:335–344
106. Tiegs SL, Russell DM, Nemazee D (1993) Receptor editing in self-reactive bone marrow B cells. J Exp Med 177:1009–1020
107. Treiber T, Mandel EM, Pott S, Gyory I, Firner S, Liu ET, Grosschedl R (2010) Early B cell factor 1 regulates B cell gene networks by activation, repression, and transcription- independent poising of chromatin. Immunity 32:714–725
108. Tsukada S, Saffran DC, Rawlings DJ, Parolini O, Allen RC, Klisak I, Sparkes RS, Kubagawa H, Mohandas T, Quan S et al (1993) Deficient expression of a B cell cytoplasmic tyrosine kinase in human X-linked agammaglobulinemia. Cell 72:279–290
109. Ubelhart R, Bach MP, Eschbach C, Wossning T, Reth M, Jumaa H (2010) N-linked glycosylation selectively regulates autonomous precursor BCR function. Nat Immunol 11:759–765
110. Ubelhart R, Hug E, Bach MP, Wossning T, Duhren-von Minden M, Horn AH, Tsiantoulas D, Kometani K, Kurosaki T, Binder CJ, Sticht H, Nitschke L, Reth M, Jumaa H (2015) Responsiveness of B cells is regulated by the hinge region of IgD. Nat Immunol 16:534–543
111. van Zelm MC, van der Burg M, de Ridder D, Barendregt BH, de Haas EF, Reinders MJ, Lankester AC, Revesz T, Staal FJ, van Dongen JJ (2005) Ig gene rearrangement steps are initiated in early human precursor B cell subsets and correlate with specific transcription factor expression. J Immunol 175:5912–5922
112. Verkoczy L, Ait-Azzouzene D, Skog P, Martensson A, Lang J, Duong B, Nemazee D (2005) A role for nuclear factor kappa B/rel transcription factors in the regulation of the recombinase activator genes. Immunity 22:519–531
113. Vetrie D, Vorechovsky I, Sideras P, Holland J, Davies A, Flinter F, Hammarstrom L, Kinnon C, Levinsky R, Bobrow M et al (1993) The gene involved in X-linked agammaglobulinaemia is a member of the src family of protein-tyrosine kinases. Nature 361:226–233

114. Villa A, Santagata S, Bozzi F, Giliani S, Frattini A, Imberti L, Gatta LB, Ochs HD, Schwarz K, Notarangelo LD et al (1998) Partial V(D)J recombination activity leads to Omenn syndrome. Cell 93:885–896
115. Vossenkamper A, Blair PA, Safinia N, Fraser LD, Das L, Sanders TJ, Stagg AJ, Sanderson JD, Taylor K, Chang F et al (2013) A role for gut-associated lymphoid tissue in shaping the human B cell repertoire. J Exp Med 210:1665–1674
116. Wang YH, Nomura J, Faye-Petersen OM, Cooper MD (1998) Surrogate light chain production during B cell differentiation: differential intracellular versus cell surface expression. J Immunol 161:1132–1139
117. Wesemann DR, Portuguese AJ, Meyers RM, Gallagher MP, Cluff-Jones K, Magee JM, Panchakshari RA, Rodig SJ, Kepler TB, Alt FW (2013) Microbial colonization influences early B-lineage development in the gut lamina propria. Nature 501:112–115
118. Wienands J, Schweikert J, Wollscheid B, Jumaa H, Nielsen PJ, Reth M (1998) SLP-65: a new signaling component in B lymphocytes which requires expression of the antigen receptor for phosphorylation. J Exp Med 188:791–795
119. Winter DR, Amit I (2014) The role of chromatin dynamics in immune cell development. Immunol Rev 261:9–22
120. Xu S, Huo J, Chew WK, Hikida M, Kurosaki T, Lam KP (2006) Phospholipase Cgamma2 dosage is critical for B cell development in the absence of adaptor protein BLNK. J Immunol 176:4690–4698
121. Zikherman J, Doan K, Parameswaran R, Raschke W, Weiss A (2012) Quantitative differences in CD45 expression unmask functions for CD45 in B-cell development, tolerance, and survival. Proc Natl Acad Sci U S A 109:E3–E12

Agammaglobulinemias: Basic Pathogenesis and Clinical Spectrum

2

Alessandro Plebani and Vassilios Lougaris

2.1 X-Linked Agammaglobulinemia (XLA)

2.1.1 Definition

X-linked agammaglobulinemia (XLA) is a rare form of primary immunodeficiency characterized by absence of circulating B cells with severe reduction in all serum immunoglobulin levels due to mutations in the gene encoding BTK (Bruton's tyrosine kinase). The incidence of the disease varies from 1:100.000 to 1:200.000 depending on ethnicity.

2.1.2 Etiology

The first patient affected with agammaglobulinemia was described by Colonel Bruton in 1952 [1]; however, it was only after four decades that the underlying genetic defect was identified [2, 3]. The gene responsible for XLA maps on the X chromosome and encodes for a member of the Tec family of kinases, i.e., Bruton's tyrosine kinase (BTK). BTK was found to be mutated in the majority of male patients [4–7]. B cells express an important receptor complex on their cell surface: the B-cell receptor (BCR) or pre-BCR in the bone marrow [8]; BTK is an important downstream kinase of the BCR-signaling cascade. The knockout mouse model for XLA (xid mouse) was particularly helpful in studying the role of BTK [9]. Normally, early B-cell development takes place in the bone marrow, where, starting from the pluripotent stem cells, B cells undergo several steps of differentiation: from pro-B

A. Plebani, MD, PhD (✉) • V. Lougaris, MD
Pediatrics Clinic, Department of Clinical and Experimental Sciences,
University of Brescia and Spedali Civili of Brescia,
P.le Spedali Civili no 1, 25123 Brescia, Italy
e-mail: alessandro.plebani@unibs.it; vlougarisbs@yahoo.com

A. Plebani, V. Lougaris (eds.), *Agammaglobulinemia*,
Rare Diseases of the Immune System 4, DOI 10.1007/978-3-319-22714-6_2

to pre-B to immature B cells to mature B cells, ready to enter the periphery [10–12]. Bone marrow analysis of patients mutated in BTK showed a consistent block at the pro-B to pre-B stage of B-cell development, underlying the important role for BTK in early B-cell development [13]. In fact, affected patients typically have less than 1–2 % of B cell in the periphery, confirming the almost complete block of early B-cell development in this disorder [14, 15]. Consequently, immunoglobulin serum levels are severely reduced, and there is no humoral response to vaccinations. The lack of B cells results in a significant reduction in size of tissues such as lymph nodes and tonsils, normally highly populated by B cells.

Mutations in BTK are transmitted in an X-linked manner and may be familiar (in this case, mothers of affected patients are healthy carriers) or de novo ones.

2.1.3 Clinical Manifestations

Maternal IgGs, transferred through the placenta, play a protective role against infections in infants for the first 6–12 months of life. In XLA patients, clinical manifestations become evident around this period, when maternal IgGs are catabolized and their protective effect is not present anymore. The hallmarks of the disease are recurrent bacterial respiratory and/or gastrointestinal infections, although some patients may remain asymptomatic for the first years of life. In rare cases, diagnosis of XLA is made in adolescence or even in adulthood due to lack of symptoms until that age. The typical range of infections in XLA patients includes recurrent otitis media, sinusitis, bronchitis, pneumonia, and gastrointestinal infections. Although the incidence of these types of infections varies among the different cohorts so far reported, it appears that the main type of infections involves the upper and lower respiratory tract [4, 6, 16, 17].

Encapsulated pyogenic bacteria are the main infectious agents in XLA, both at diagnosis and upon immunoglobulin replacement treatment. Bacteria such as *Streptococcus pneumoniae*, *Haemophilus influenzae*, *Staphylococcus aureus*, and others are most frequently identified as causative in these patients (Table 2.1). In the case of invasive bacterial infections, such as septicemia, the main infectious agents are *Pseudomonas species*, followed by *H. influenzae*, *S. pneumoniae*, and *S. aureus*. Bacterial meningitis on the other hand may complicate the clinical history of these patients, especially before diagnosis, and is caused by the same infectious agents mentioned before. Before diagnosis, septic arthritis may complicate the clinical history of affected patients and is mainly caused by *H. influenzae* and *S. pneumoniae*, while once immunoglobulin replacement treatment is initiated, viral agents are frequently responsible.

Recurrent infections of the upper and lower respiratory tract are an important clinical problem for patients with XLA, both before and after diagnosis [4, 6, 16]. Chronic sinusitis is reported to be a consistent clinical finding in almost two thirds of affected patients, even upon Ig treatment. Similarly, recurrent infections of the lower respiratory tract, even under Ig replacement treatment, lead to the development of bronchiectasis compromising severely the clinical course of the disease (Fig. 2.1).

Table 2.1 Most frequently isolated bacterial infectious agents in X-linked agammaglobulinemia according to the site of infections

Etiology of pneumonia in XLA	
Haemophilus influenzae type b	58 %
Streptococcus pneumoniae	17 %
Staphylococcus aureus	17 %
Pseudomonas aeruginosa	8 %
Etiology of sinusitis in XLA	
Haemophilus influenzae type b	67 %
Streptococcus pneumoniae	14 %
Staphylococcus aureus	10 %
Klebsiella	3 %
Moraxella	3 %
Pseudomonas aeruginosa	2 %
Etiology of GI infections in XLA	
Giardia	60 %
Salmonella	20 %
Campylobacter	8 %
Escherichia coli	8 %
Blastocystis	4 %

Data based on long-term follow-up of 125 XLA patients, courtesy of the IPINET registry, Italy

Infections of the gastrointestinal tract represent an important clinical burden in XLA (Table 2.1). One of the most frequently isolated infectious agents is *Giardia lamblia*, and unfortunately, its eradication is not always successful, leading to protracted and/or recurrent diarrhea and malabsorption (Fig. 2.2). A similar clinical picture may be caused by *Campylobacter jejuni*, and several cases of gastrointestinal infections caused by *Salmonella* have also been reported in XLA. *Helicobacter-like* organisms may cause invasive infections with fever and lower limb cellulitis that may progress to a "woody-appearing" skin lesion (Fig. 2.3) and require long period of intravenous antibiotics. Finally, *Helicobacter cinaedi* bacteremia was recently reported in an adult XLA patient presenting with macules but without fever [18].

Besides the infectious agents mentioned so far, *Mycoplasma species* may cause infections in XLA patients, especially of the respiratory and urogenital tract and in rare cases the joints. It is not always easy to isolate the infectious agent, and therefore, the clinical course may be prolonged and severe. *Mycoplasma* may coexist with other bacterial agents during infections rendering the disease more severe. Bacterial conjunctivitis is relatively frequent in agammaglobulinemia, affecting almost 6–8 % of patients.

Besides bacteria, also virus may complicate the clinical history of patients affected with XLA. The most frequently isolated virus in XLA is enteroviruses, namely, poliovirus, echovirus, and coxsackievirus. Patients with XLA have been reported to develop vaccine-associated poliomyelitis in the case of vaccination with the live attenuated oral vaccine (Sabin) associated with a high mortality rate.

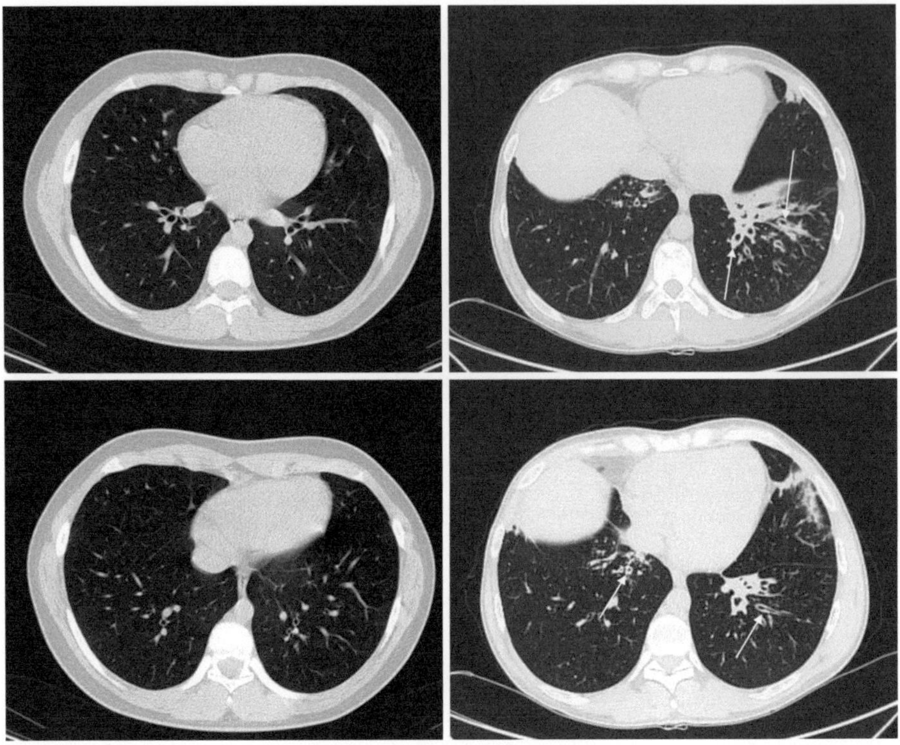

Fig. 2.1 Lung CT scans from XLA patients. While some patients do not develop bronchiectasis (*left panels*), many XLA patients tend to develop bronchiectasis (*right panels*, *white arrows*) even when under regular immunoglobulin replacement treatment (Courtesy of Dr. MP Bondioni, Pediatric Radiology Unit, University of Brescia, Spedali Civili di Brescia, Italy)

Besides the classical manifestations associated with the onset of the enteroviral infection, XLA patients may also develop subacute progressive neurological symptoms such as ataxia, loss of cognitive skills, paresthesias, and sensorineural hearing loss; these symptoms should always raise the suspicion of an infection caused by enterovirus. In fact, XLA patients may develop enteroviral meningoencephalitis mainly with a slow, chronic, and progressive pattern, although acute onset with fever and seizures has also been reported in a limited number of cases. It is not always easy to isolate the enterovirus from the CSF. In recent years, the application of molecular biology techniques such PCR was initially thought to be able to overcome this problem; however, despite this technical improvement, enteroviral detection is not always successful. Typical findings in the CSF of these patients may include pleocytosis, elevated protein content, and in some cases hypoglycorrhachia, suggestive of a viral infection, although most frequently the CSF characteristics in XLA patients may be almost normal. IVIG treatment radically reduced the incidence of chronic enteroviral infections in XLA patients, although they are not completely eradicated yet. In some cases, the protective effect of IVIG against enteroviral

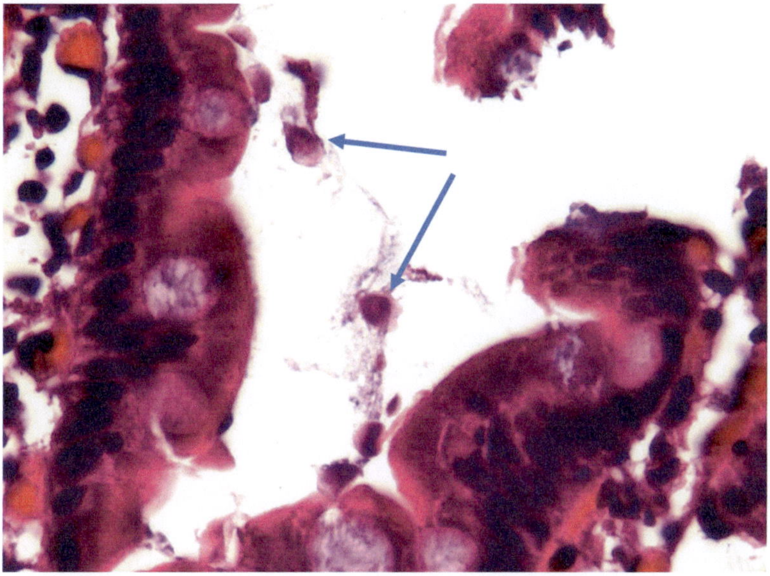

Fig. 2.2 Giardia lamblia gastrointestinal infection in XLA. Histological examination of gastric biopsy specimen from an XLA patient showing the presence of Giardia lamblia (*blue arrows*) (Courtesy of Dr. V. Villanacci, Institute of Pathology, Spedali Civili, Brescia, Italy)

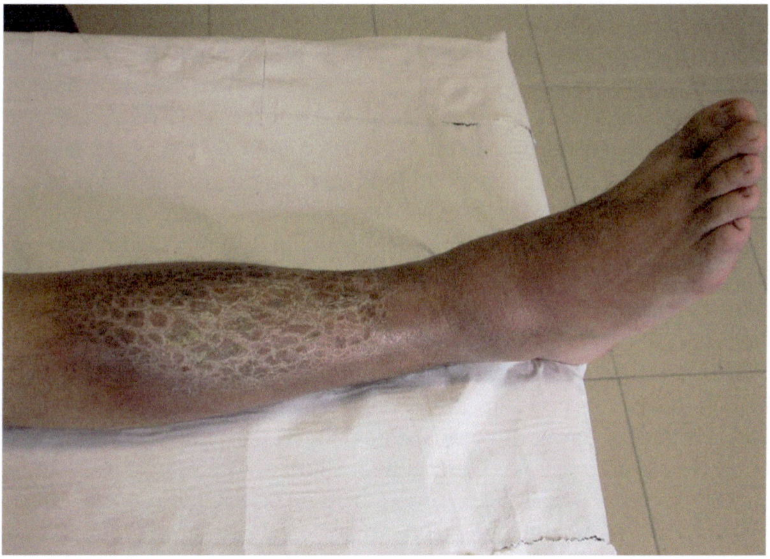

Fig. 2.3 Skin manifestations during Helicobacter-like organism infection in an XLA patient. Erythemato-squamous lesions of the lower right art during a systemic infection sustained by Helicobacter-like species

infections in XLA has been associated to high-dose treatment (and not the usual dosage applied in replacement treatment), although the limited number of patients reported does not allow definitive conclusions. The mechanism proposed is based on the anti-inflammatory effect of high-dose immunoglobulins, although further data are needed to validate the effective mechanism. Combined treatment with high-dose immunoglobulins and pleconaril has been reported to be effective in controlling enteroviral infection in patients affected with XLA [19–21]. Although reported data have shown certain efficacy of pleconaril in this type of infection, this drug is no longer available for compassionate use. Brain imaging such as MRI or CT scans is almost always normal at the onset of the disease, while chronic enteroviral infections are progressively characterized by the development of cerebral edema, diffuse inflammation, and progressive cerebral atrophy [22–25]. Rare cases of leptomeningitis in XLA patients (instead of the classical form of encephalitis) have also been reported.

Besides neurological involvement, CNS enteroviral infection may also present with other symptoms such as peripheral edema and erythematosus rash that may resemble a dermatomyositis-like syndrome [26]. When biopsies are performed from the skin or the muscle, they evidence inflammation as main feature. Enteroviral infection may also present with liver involvement including ALT increase and hepatomegaly, with or without fever or rash. The peripheral involvement described here is characteristic of disseminated enteroviral infection, which is not always characterized by a favorable prognosis.

Immunoglobulin preparations are quite safe nowadays; however, in the early 1990s, hepatitis C contamination was reported in XLA patients. Interestingly, it seems that XLA patients tolerate better this infection when compared to patients affected with CVID. Among reported XLA patients with hepatitis C, more than one third of infected patients cleared the infection or remained asymptomatic, and only one patient developed hepatic failure, but he was also coinfected with hepatitis B. Rare manifestations in XLA patients include pneumonia caused by *Pneumocystis jiroveci* (Fig. 2.4) [27–29], recurrent pyoderma (being the only clinical finding in an XLA patient) [30], and chronic gingivitis (being the only clinical finding in an XLA patient) [31].

Joint involvement may be present in XLA patients in almost 20 % of cases [4, 6, 16, 17]. The typical clinical presentation resembles that of rheumatoid arthritis (RA) with pain, motion limitation, effusion, and destructive pannus formation. Although in most cases no isolates are found, a pyogenic cause may be identified in a limited number of cases. Empirical treatment consists of IVIG treatment, with frequently beneficial effect, in addition with antibiotic treatment, suggesting a possible infectious cause. Among the reported isolates, enterovirus and *Mycoplasma species* are the most frequent ones. Although B cells have been reported to be associated with the pathogenesis of RA, no B-cell infiltrates were found in the synovium of XLA patients with RA.

The hematologic manifestation that has been mainly reported in XLA is neutropenia. The percentage of XLA patients with neutropenia, mainly of secondary nature, before IVIG treatment is variable, depending on the cohort of patients

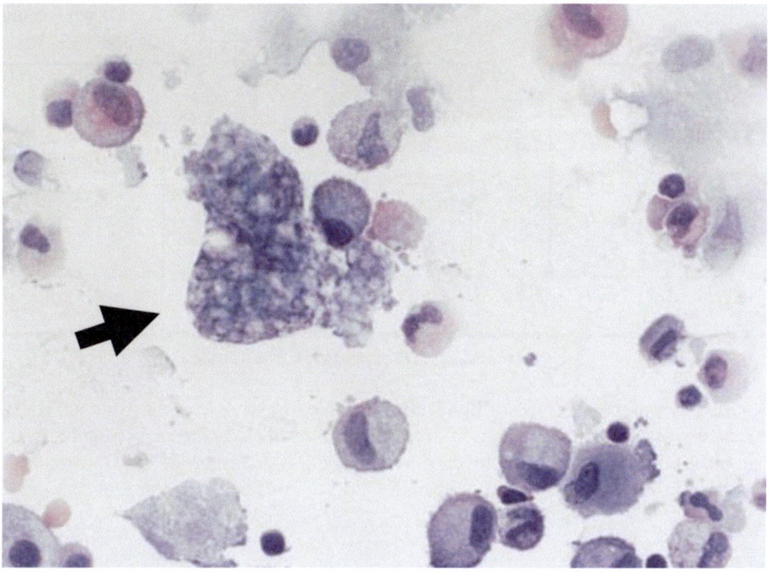

Fig. 2.4 Bronchoalveolar lavage from an XLA patient showing *Pneumocystis jiroveci*. A patient affected with XLA and respiratory symptoms nonresponsive to classical antibiotic treatment underwent bronchoalveolar lavage revealing the presence of *Pneumocystis jiroveci* (*black arrow*). This patient was under steroid therapy

reported, ranging from 10 to 25 % [32–36]. However, to date, the precise role of BTK in neutrophil development/homeostasis is not completely clear yet. XLA patients have also been reported to develop acute lymphoblastic leukemia [37].

Finally, other manifestations that have been reported in XLA patients include glomerulonephritis [38], alopecia, amyloidosis, and von Recklinghausen disease [39].

2.1.4 Diagnosis

The hallmark of XLA is the absence of peripheral B cells (<2 %) in the presence of very low to absent immunoglobulin serum levels of all classes [40, 41]. In rare cases, B cells may be detected in the periphery, but additional immunological workup such as recall antibody responses to specific antigens may be of additional help. Once the clinical suspicion is confirmed from the immunological examination, BTK protein expression in cells other than B cells such as monocytes can be a rapid and economic diagnostic tool allowing both for early diagnosis of affected patients and identification of healthy carriers (Fig. 2.5). The genetic analysis should always follow BTK expression testing, being the former considered the gold standard for a definite diagnosis of XLA. In the case a mutation in BTK is found in the affected patient, the mother's carrier status should be examined. Finally, prenatal diagnosis can also be performed once the mutation in BTK is identified.

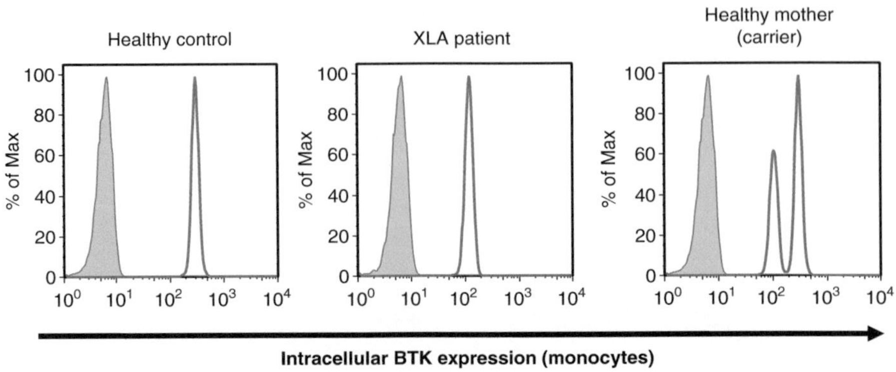

Fig. 2.5 Flow cytometric analysis of BTK intracellular expression in monocytes. BTK expression in human monocytes shown as histogram in a healthy control (*left panel*), an affected patient (*mid panel*), and his mother who is a healthy carrier (*right panel*) (Courtesy of Dr. D. Moratto, Institute for Molecular Medicine "A. Nocivelli," Spedali Civili di Brescia, University of Brescia, Brescia, Italy)

Table 2.2 List of genes associated with autosomal recessive agammaglobulinemia (ARA)

Gene	Protein	Chromosome	OMIM
IGHM	Mu HC	14q32.33	147020
CD79A	Igα	19q13.2	112205
CD79B	Igβ	17q23.3	147245
BLNK	BLNK	10q24.1	604515
IGLL1	λ5	22q11.23	146770
PIK3R1	p85α	5q13.1	171833

2.2 AR-Agammaglobulinemia (MUHC, IGA, IGB, L5, BLNK, and p85α)

2.2.1 Definition

Autosomal recessive agammaglobulinemia (ARA) affects both males (in the absence of BTK mutations) and females and is a rare form of primary immunodeficiency characterized by severe reduction of all immunoglobulin classes and absence of peripheral B cells [40, 41]. The underlying genetic defect is currently known only in a limited number of patients (Table 2.2).

2.2.2 Etiology

Early B-cell development takes place in the bone marrow through a highly orchestrated process with the sequential and highly regulated expression of specific gene products that regulate the maturation from pro-B to pre-B to immature B cells to mature B cells that can entry the periphery [42–47]. An important step in this differentiation process is the formation of the pre-B cell that expresses on the cell surface the pre-BCR (pre-B-cell receptor) complex formed by the mu heavy chain, Igα, Igβ, VpreB, and λ5 [8]. Pre-BCR is essential for the signaling events that promote further differentiation of B cells trough kinases such as BTK, BLNK, and others. Furthermore, kinases dependent on other signaling cascades, such as PI3K (e.g., downstream of the CD19-CD21-CD81 complex), have a positive effect on the BCR-signaling cascade [48, 49]. The important role of these molecules in early B-cell development was demonstrated through both animal models and in vitro models, rendering therefore the genes encoding for these proteins' candidate genes for agammaglobulinemia of unknown genetic origin.

In 1996, the first patients mutated in mu heavy chain with autosomal recessive agammaglobulinemia (ARA) were described [50]. Two parallel studies in the USA and in Italy studied the prevalence of mu heavy chain deficiency among patients affected with ARA which reached around 40–50 % among the studied patients [51].

The surrogate light chains λ5/14.1 and VpreB are part of the pre-BCR complex and are important of the early B-cell development. A single patient with mutations in the λ5/14.1 gene has been reported so far [52].

The two signaling transducing molecules Igα and Igβ associate with the pre-BCR and the BCR and are essential for B-cell development [8, 10, 11]. So far, five patients have been reported to be affected with Igα deficiency. The mutation in Igα in the first patient identified resulted in alternative exon splicing of the gene product which abolishes the expression of the protein on the cell surface [53]. The second reported patient, of Turkish origin, carried a homozygous alteration at an invariant splice donor site of intron 2 [54]. Two additional patients with defects in the extracellular domain were also reported [15]. Finally, the fifth reported patient, of Iranian origin, carried a novel homozygous nonsense mutation in the gene encoding Igα [55].

To date, only three patients with biallelic mutations in Igβ have been reported. The first one is a female patient carrying a hypomorphic mutation in Igβ and a leaky defect in B-cell development [56]. The second patient, a male of 20 years of age, presented a homozygous nonsense mutation resulting in a stop codon, resulting in a complete block of B-cell development at the pro-B- to pre-B-cell transition in the bone marrow, similarly to what observed in the animal model [57]. Finally, the third patient is a female child that carries a novel homozygous nonsense mutation in the gene encoding for Igβ [58].

Upon BCR cross-linking, various kinases are involved in the downstream signaling cascade. One of these, BLNK (also called SLP-65), was initially found to be mutated in a single patient with a specific block at the pro-B to pre-B stage in the bone marrow [59]. Recent experimental data have underlined the important role of BLNK for B-cell homeostasis and differentiation upon BCR expression on the cell surface [60]. Other two siblings with deleterious frameshift mutations in BLNK and agammaglobulinemia were recently reported [61].

PI3K comprises a family of kinases expressed in various cell types that play important roles in various biological processes such as cell cycle, metabolism cell growth, migration, and others. Regarding B cells, a novel homozygous nonsense mutation in p85α was recently identified in a single female patient with agammaglobulinemia, leading to an earlier block in B-cell development in the bone marrow, before the expression of CD19 on the surface of B cells [62]. Interestingly, although expressed in other cell types such as T cells, dendritic cells, and others, these appear functionally unaffected from the presence of this mutation.

2.2.3 Clinical Manifestations

Clinical findings in patients affected with ARA resemble those of XLA, although apparently in a more severe manner. Frequently, the onset of the disease is also earlier.

Regarding mu heavy chain deficiency, the infectious spectrum at onset is rather similar to that observed in XLA: recurrent otitis media, bronchitis, sinusitis, pneumonia, chronic enteroviral encephalitis, and sepsis (frequently caused by *Pseudomonas aeruginosa*) [50, 51]. As observed in XLA, the clinical picture ameliorated upon Ig replacement treatment is initiated on a regular basis. Chronic intestinal infection sustained by *Giardia lamblia*, resistant to treatment, has been observed in one female patient (Plebani, personal communication). Neutropenia is a rather frequent hematological finding in this disorder, since it may be observed in almost one third of patients and usually normalizes upon immunoglobulin replacement therapy.

The limited number of patients affected with the other reported autosomal recessive defects leading to agammaglobulinemia does not allow for conclusive remarks. The only patient reported so far to be affected with λ5/14.1 deficiency presented an episode of *Haemophilus* meningitis and recurrent otitis media and was found to be hypogammaglobulinemic at the age of 5 years. Peripheral B cells were almost undetectable (<0.06 %). Bone marrow analysis confirmed the early block in B-cell development at the pro-B to pre-B stage.

The clinical spectrum of patients affected with Igα deficiency is variable, even though a limited number of patients (five) have been reported so far. The first reported patient presented an early onset with diarrhea and failure to thrive in the first month of life [53]. She was admitted at the age of 12 months for bronchitis and neutropenia. Immunological workup showed absence of peripheral B cells and undetectable serum immunoglobulins; the patients did not have detectable lymph nodes. Bone marrow analysis showed a pro-B to pre-B block in the early steps of B-cell development. The second reported patient (male) presented a clinical history of diarrhea and respiratory infections; he also developed a dermatomyositis-like phenotype. Unfortunately, the patient died of a pulmonary infection [54]. The female patient carrying the early stop codon in Igα presented a severe neurological manifestation resembling initially a febrile seizure that was however not responsive to classical therapeutic approaches [55]. The continuous worsening of the clinical status of the patient led to the immunological workup showing agammaglobulinemia and absence of peripheral B cells. Clinical data on the other two patients affected with Igα deficiency are not available.

Three patients affected with Igβ deficiency have been reported so far. The patient carrying the hypomorphic mutation had a history of recurrent lower respiratory infections starting from 5 months of age, while diagnosis was made at the age of 15 months [56]. Immunoglobulin replacement treatment ameliorated the patient's clinical condition. The patient with the first nonsense mutation reported in Igβ was admitted at the age of 8 months for pneumonia and salmonella enteritis [57]. During this admission, his immunological workup showed indosable immunoglobulin serum levels for all classes and absence of peripheral B cells (<1 %). Although the patient was started on IVIG, his clinical history was complicated by recurrent respiratory infections both of the upper and the lower respiratory tract and bacterial conjunctivitis. The third patient carrying a novel homozygous mutation in Igβ presented a history of respiratory infections; at the age of 15 months, she was admitted due to echthyma of the left gluteus and neutropenia [58]. Immunological workup showed lack of serum immunoglobulins of all classes and absence of peripheral B cells. IVIG treatment was started and the neutropenia resolved within three months.

The clinical presentation of the first patient affected with BLNK deficiency included, by the age of 8 months, two episodes of pneumonia and recurrent otitis [59]. Once the diagnosis of agammaglobulinemia was made due to low immunoglobulin serum levels and absence of peripheral B cells, he was started on IVIG; however, he continued presenting recurrent otitis media and sinusitis and developed protein-losing enteropathy during adolescence. Unfortunately, he also developed hepatitis C related to the immunoglobulin preparations. Another two patients (siblings) affected with BLNK deficiency were recently reported [61]. The male patient was diagnosed at the age of 6 months with a clinical history of recurrent otitis and diarrhea [61]. During follow-up he developed arthritis and diffuse dermatitis. PCR revealed the presence of enteroviral infection in peripheral blood, although the skin biopsy did not evidence the presence of the virus. The female sister of the above-mentioned patient was diagnosed at the age of 12 months with a clinical history of recurrent otitis and sinopulmonary infections. Her clinical history was complicated by diarrhea, obstructive lung disease, and arthralgia [61].

The female patient affected with p85a deficiency presented an early onset of symptoms: at the age of 3.5 months, she was admitted for neutropenia, gastroenteritis, and interstitial pneumonia. Immunological workup showed agammaglobulinemia in the absence of peripheral B cells [62]. During follow-up and into her adolescent years, the patient developed erythema nodosum, arthritis, *Campylobacter* bacteremia, and inflammatory bowel disease. No metabolic alteration was reported.

2.2.4 Diagnosis

The presentation of ARA is similar to that of XLA: low to undetectable immunoglobulin serum levels in the almost complete absence of peripheral B cells, as defined by CD19 and CD20 expression (<2 %). Male patients, once BTK deficiency is excluded, and affected female patients should undertake sequencing analysis for mu heavy chain, Igα, Igβ, λ5, BLNK, and p85α for a definite diagnosis.

2.3 Autosomal Dominant Agammaglobulinemia (E47/TCF3)

Until recently, X-linked (the more frequent) or autosomal recessive forms of agammaglobulinemia were known. However, autosomal dominant E47/TCF3 deficiency was recently described in four patients with agammaglobulinemia and reduced peripheral B cells that expressed CD19 but lacked BCR expression on the cell surface [63].

The broadly expressed transcription factor E47 resulted mutated in the four patients with TCF3 deficiency. The role of E47 in B-cell development had been previously highlighted in the animal model [64]. All reported patients harbored the heterozygous E555K mutation and showed an unusual peripheral B-cell phenotype: enhanced CD19 expression with absent expression of BCR. B-cell development in the bone marrow was performed in two out of four affected patients and showed a block of B-cell development earlier than that observed in XLA or ARA. Affected patients presented a complicated clinical history compatible with agammaglobulinemia: pneumococcal meningitis, recurrent otitis, vaccine-associated poliomyelitis, and arthritis. Associated clinical features included eosinophilic dermatitis and hepatomegaly [63].

2.4 Management of Agammaglobulinemia (X-Linked, Autosomal Recessive, and Autosomal Dominant Forms)

Agammaglobulinemia is a humoral immunodeficiency, and as such, patients' management is based on immunoglobulin replacement treatment. The introduction of immunoglobulin replacement treatment has radically changed the prognosis of affected patients. In the past, the intramuscular route of administration was applied with the major pitfall of not reaching protective IgG trough serum levels. Current routes of administration are two: intravenous or subcutaneous. While intravenous preparations are administered every 21–28 days at a dose of 400 mg/kg/dose, the subcutaneous ones are administered weekly at a dose of 100 mg/kg/dose. The main objective of the replacement treatment has been to maintain pre-infusion IgG levels >500 mg/dl. However, in recent years, it is becoming more and more evident that the dose should be more patient oriented in order to obtain the maximum benefit.

Considering the rarity of autosomal recessive and even more autosomal dominant forms of agammaglobulinemia, the majority of available data in the literature is based on XLA. In fact, long-term follow-up studies in patients affected with XLA have demonstrated that they tend to develop lung complications (chronic lung disease, CLD), even if Ig replacement treatment is performed regularly. This may depend on different factors: delayed diagnosis, previous intramuscular route of administration (reduced IgG availability and therefore lower trough levels), secreted antibodies cannot be substituted (Ig preparations contain only IgG), and Ig preparations contain pools of poly-specific IgGs, nonselected on antigen specificity. One of the major factors affecting XLA patients' prognosis is the development of CLD [6].

Since Ig replacement treatment alone is not always sufficient to prevent the development of bronchiectasis and consequently CLD, respiratory physiotherapy may have a beneficial effect on long-term outcome in XLA patients in terms of maintenance and even improvement of lung function.

Infectious episodes in patients affected with agammaglobulinemia should always be treated with antibiotics. In some cases, affected patients under Ig replacement treatment may require brief or long periods of antibiotic prophylaxis, depending on their clinical conditions. Considering the monogenic defect in XLA, novel therapeutic approaches have been proposed in recent years. Gene therapy is one of them and has been applied in murine models with encouraging results [65], although clinical application is not under consideration yet. Another recent therapeutic approach used antisense oligonucleotides (ASOs), compounds that have the ability to modulate pre-mRNA splicing and alter gene expression, in BTK mutations that affect normal mRNA splicing [66]. Further experiments are required before considering this option in patients' management.

As mentioned before, and although regular Ig replacement treatment and correct antibiotic therapy have radically modified patients' clinical history, their prognosis is still conditioned by the occurrence of complications. In particular, pulmonary complications such as development of bronchiectasis and CLD still occur, and so far, it appears that the right therapeutic approach is still to be defined [6].

Regarding ARA, the clinical data regarding follow-up and clinical complications is rather limited, due to the small number of reported cases. Nonetheless, so far, the prognosis for ARA appears similar to that of XLA.

Finally, XLA patients are reported to have an increased incidence of malignancies, including colorectal cancer, gastric adenocarcinoma, and lymphoid malignancies [4, 6, 16, 17, 37, 67–69].

References

1. Bruton OC (1952) Agammaglobulinemia. Pediatrics 9:722–728
2. Tsukada S, Saffran DC, Rawlings DJ, Parolini O, Allen RC, Klisak I, Sparkes RS, Kubagawa H, Mohandas T, Quan S (1993) Deficient expression of a B cell cytoplasmic tyrosine kinase in human X-linked agammaglobulinemia. Cell 72:279–280
3. Vetrie D, Vorechovski I, Sideras P, Holland J, Davies A, Flinter F, Hammarstrom L, Kinnon C, Levinsky R, Bobrow M (1993) The gene involved in X-linked agammaglobulinemia is a member of the src family of protein-tyrosine kinases. Nature 361:226–233
4. Lederman HM, Winkelstein JA (1985) X-linked agammaglobulinemia: an analysis of 96 patients. Medicine (Baltimore) 64:145–156
5. Ochs HD, Smith CI (1996) X-linked agammaglobulinemia. A clinical and molecular analysis. Medicine (Baltimore) 75(6):287–299, Review
6. Plebani A, Soresina A, Rondelli R, Amato G, Assari C, Cardinale F, Cazzola G, Consolini R, De Mattia D, Dell'Erba G (2002) Clinical, immunological and molecular analysis of a large cohort of patients with X-linked agammaglobulinemia: an Italian multicenter study. Clin Immunol 104:221–230
7. Vihinen M, Mattsson PT, Smith CI (1997) BTK, the tyrosine kinase affected in X-linked agammaglobulinemia. Front Biosci 2:d27–d42, Review

8. Clark MR, Cooper AB, Wang LD, Aifantis I (2005) The pre-B cell receptor in B cell development: recent advances, persistent questions and conserved mechanisms. Curr Top Microbiol Immunol 290:87–103, Review
9. Rawlings DJ, Safran DC, Tsukada S et al (1993) Mutation of unique region of Bruton's tyrosine kinase in immunodeficient XID mice. Science 261(5119):358–361
10. Yankee TM, Clark EA (2000) Signaling through the B cell antigen receptor in developing B cells. Rev Immunogenet 2(2):185–203, Review
11. Zhang M, Srivastava G, Lu L (2004) The pre-B cell receptor and its function during B cell development. Cell Mol Immunol 1(2):89–94, Review
12. Cancro MP (2004) Peripheral B-cell maturation: the intersection of selection and homeostasis. Immunol Rev 197:89–101, Review
13. Tsukada S, Rawlings DJ, Witte ON (1994) Role of Bruton's tyrosine kinase in immunodeficiency. Curr Opin Immunol 6(4):623–630, Review
14. Conley ME, Broides A, Hernandes-Trujillo V, Howard V, Kanegane H, Miywaki T, Shurtleff SA (2005) Genetic analysis of patients with defects in early B cell development. Immunol Rev 203:216–234
15. Conley ME (2002) Early defects in B cell development. Curr Opin Allergy Clin Immunol 2(6):517–522, Review
16. Moin M, Aghamohammadi A, Farhoudi A, Pourpak Z, Resaei N, Movahedi M, Gharagozlou M, Ghazi BM, Zahed A, Abolmaali K, Mahmoudi M, Amami L, Bashashati M (2004) X-linked agammaglobulinemia: a survey of 33 Iranian patients. Immunol Invest 33:81–93
17. Stewart DM, Lian L, Nelson DL (2001) The clinical spectrum of Bruton's agammaglobulinemia. Curr Allergy Asthma Rep 1(6):558–565, Review
18. Simons E, Spacek LA, Lederman HM, Winkelstein JA (2004) Helicobacter cinaedi bacteremia presenting as macules in an afebrile patient with X-linked agammaglobulinemia. Infection 32(6):367–368
19. Wildenbeest JG, van den Broek PJ, Benschop KS et al (2012) Pleconaril revisited: clinical course of chronic enteroviral meningoencephalitis after treatment correlates with in vitro susceptibility. Antivir Ther 17(3):459–466
20. van de Ven AA, Douma JW, Rademaker C et al (2011) Pleconaril-resistant chronic parechovirus-associated enteropathy in agammaglobulinemia. Antivir Ther 16(4):611–614
21. Schmugge M, Lauener R, Bossart W et al (1999) Chronic enteroviral meningo-encephalitis in X-linked agammaglobulinemia: favourable response to anti-enteroviral treatment. Eur J Pediatr 158(12):1010–1011
22. Halliday E, Winkelstein JA, Webster AD (2003) Enteroviral infections in primary immunodeficiency: a survey of morbidity and mortality. J Infect 46:1–8
23. Misbah SA, Spickett GP, Ryba PC, Hockaday JM, Kroll JS, Sherwood C, Kurtz JB, Moxon ER, Chapel HM (1992) Chronic enteroviral meningoencephalitis in agammaglobulinemia: case report and literature review. J Clin Immunol 12:266–270
24. Sarpong S, Skolnick HS, Ochs HD, Futatani T, Winkelstein JA (2002) Survival of wild polio by a patient with X-linked agammaglobulinemia. Ann Allergy Asthma Immunol 88:59–63
25. Wilfert CM, Buckley RH, Mohanakumar T, Griffith JF, Katz SL, Whisnant JK, Eggleston PA, Moore M, Treadwell E, Oxman MN, Rosen FS (1977) Persistent and fatal central nervous system ECHOvirus infections in patients with agammaglobulinemia. N Engl J Med 296: 1485–1489
26. Bardelas JA, Winkelstein JA, Seto DS, Tsai T, Roqol AD (1977) Fatal ECHO 24 infection in a patient with hypogammaglobulinemia: relationship to dermatomyositis-like syndrome. J Pediatr 90:396–404
27. Alibrahim A, Lepore M, Lierl M, Filipovich A, Assaad A (1998) Pneumocystis carinii pneumonia in an infant with X-linked agammaglobulinemia. J Allergy Clin Immunol 101:552–553
28. Dittrich AM, Schulse I, Magdorf K, Wahn V, Wahn U (2003) X-linked agammaglobulinaemia and Pneumocystis carinii pneumonia-an unusual coincidence. Eur J Pediatr 162:432–433
29. Saulsbury FT, Bernstein MT, Winkelstein JA (1979) Pneumocystis carinii pneumonia as the presenting infection in congenital hypogammaglobulinemia. J Pediatr 95:559–565

30. Lin MT, Chien YH, Shyur SD, Huang LH, Chiang YC, Wen DC, Liang PH, Yang HC (2006) De novo mutation in the BTK gene of atypical X-linked agammaglobulinemia in a patient with recurrent pyoderma. Ann Allergy Asthma Immunol 96(5):744–748
31. Liu AJ, Dao-Ung LP, McDonald D, Nanan R (2006) Chronic gingivitis in a new BTK mutation. Eur J Haematol 76(2):171–175
32. Buckley RH, Rolands DTJ (1973) Agammaglobulinemia, neutropenia, fever, and abdominal pain. J Allergy Clin Immunol 51:308–318
33. Cham B, Bonilla MA, Winkelstein JA (2002) Neutropenia associated with primary immunodeficiency diseases. Semin Hematol 39:107–115
34. Farrar JE, Rohrer J, Conley ME (1996) Neutropenia in X-linked agammaglobulinemia. Clin Immunol Immunopathol 81:271–276
35. Kanegane H, Taneichi H, Nomura K, Futatani T, Miyawaki T (2005) Severe neutropenia in Japanese patients with X-linked agammaglobulinemia. J Clin Immunol 25:491–495
36. Plo Rodriguez F, Garcia Rodriguez M, Ferreira Cerdan A, Fontan Casariego G (1999) Neutropenia as early manifestation of X-linked agammaglobulinemia. Report on 4 patients. An Esp Pediatr 51:235–240
37. Hoshino A, Okuno Y, Migita M, Ban H, Yang X, Kiyokawa N, Adachi Y, Kojima S, Ohara O, Kanegane H (2015) X-linked agammaglobulinemia associated with B-precursor Acute Lymphoblastic Leukemia. J Clin Immunol 35(2):108–11
38. Lavrador V, Correia F, Sampaio R, Cândido C, Sameiro-Faria M, Marques L, Mota C (2014) Membranoproliferative glomerulonephritis and x-linked agammaglobulinemia: an uncommon association. Case Rep Pediatr 2014:480947
39. Hirata D, Nara H, Inaba T, Muroi R, Kanegane H, Miyawaki T, Okazaki H, Minota S (2002) von Recklinghausen disease in a patient with X-linked agammaglobulinemia. Intern Med 41(11):1039–1043
40. Al-Herz W, Bousfiha A, Casanova J-L, Chapel H, Conley ME, Cunningham-Rundles C, Etzioni A, Fischer A, Franco JL, Geha RS, Hammarström L, Nonoyama S, Notarangelo LD, Ochs HD, Puck JM, Roifman CM, Seger R, Tang ML (2011) Primary immunodeficiency diseases: an update on the classification from the International Union of Immunological Societies Expert Committee for Primary Immunodeficiency. Front Immunol 2:54
41. Conley ME, Notarangelo LD, Etzione A (1999) Diagnostic criteria for primary immunodeficiencies. Representing PAGID (Pan-American Group for Immunodeficiency) and ESID (European Society for Immunodeficiencies). Clin Immunol 93:190–197
42. Bartholdy B, Matthias P (2004) Transcriptional control of B cell development and function. Gene 327(1):1–23, Review
43. Espeli M, Rossi B, Mancini SJ, Roche P, Gauthier L, Schiff C (2006) Initiation of pre-B cell receptor signaling: common and distinctive features in human and mouse. Semin Immunol 18(1):56–66, Review
44. Hagman J, Lukin K (2006) Transcription factors drive B cell development. Curr Opin Immunol 18(2):127–134, Review
45. Lazorchak AS, Wojciechowski J, Dai M, Zhuang Y (2006) E2A promotes the survival of precursor and mature B lymphocytes. J Immunol 177(4):2495–2504
46. Lin H, Grosschedl R (1995) Failure of B-cell differentiation in mice lacking the transcription factor EBF. Nature 376(6537):263–267
47. Rudin CM, Thompson CB (1998) B-cell development and maturation. Semin Oncol 25(4):435–446, Review
48. Werner M, Hobeika E, Jumaa H (2010) Role of PI3K in the generation and survival of B cells. Immunol Rev 237(1):55–71
49. Donahue AC, Fruman DA (2004) PI3K signaling controls cell fate at many points in B lymphocyte development and activation. Semin Cell Dev Biol 15(2):183–197, Review
50. Yel L, Minegishi Y, Coustain-Smith E, Buckley RH, Trubel H, Pachman LM, Kitchingman GR, Campana D, Rohrer J, Conley ME (1996) Mutations in the mu heavy-chain gene in patients with agammaglobulinemia. N Engl J Med 335(20):1486–1493

51. Ferrari S, Zuntini R, Lougaris V, Soresina A, Sourková V, Fiorini M, Martino S, Rossi P, Pietrogrande MC, Martire B, Spadaro G, Cardinale F, Cossu F, Pierani P, Quinti I, Rossi C, Plebani A (2007) Molecular analysis of the pre-BCR complex in a large cohort of patients affected by autosomal recessive agammaglobulinemia. Genes Immun 8(4):325–333
52. Minegishi Y, Coustan-Smith E, Wang YH, Cooper MD, Campana D, Conley ME (1998) Mutations in the human lambda5/14.1 gene result in B cell deficiency and agammaglobulinemia. J Exp Med 187(1):71–77
53. Minegishi Y, Coustan-Smith E, Rapalus L, Ersoy F, Campana D, Conley ME (1999) Mutations in Igalpha (CD79a) result in a complete block in B-cell development. J Clin Invest 104(8):1115–1121
54. Wang Y, Kanegane H, Sanal O, Tezcan I, Ersoy F, Futatani T, Miyawaki T (2002) Novel Igalpha (CD79a) gene mutation in a Turkish patient with B cell-deficient agammaglobulinemia. Am J Med Genet 108(4):333–336
55. Khalili A, Plebani A, Vitali M, Abolhassani H, Lougaris V, Mirminachi B, Rezaei N, Aghamohammadi A (2014) Autosomal recessive agammaglobulinemia: a novel non-sense mutation in CD79a. J Clin Immunol 34(2):138–141
56. Dobbs AK, Yang T, Farmer D, Kager L, Parolini O, Conley ME (2007) Cutting edge: a hypomorphic mutation in Igbeta (CD79b) in a patient with immunodeficiency and a leaky defect in B cell development. J Immunol 179(4):2055–2059
57. Ferrari S, Lougaris V, Caraffi S, Zuntini R, Yang J, Soresina A, Meini A, Cazzola G, Rossi C, Reth M, Plebani A (2007) Mutations of the Igβ gene cause agammaglobulinemia in man. J Exp Med 204(9):2047–2051
58. Lougaris V, Vitali M, Baronio M, Moratto D, Tampella G, Biasini A, Badolato R, Plebani A (2014) Autosomal recessive agammaglobulinemia: the third case of Igβ deficiency due to a novel non-sense mutation. J Clin Immunol 34(4):425–427
59. Minegishi Y, Rohrer J, Coustan-Smith E, Lederman HM, Pappu R, Campana D, Chan AC, Conley ME (1999) An essential role for BLNK in human B cell development. Science 286(5446):1954–1957
60. Lagresle-Peyrou C, Millili M, Luce S, Boned A, Sadek H, Rouiller J, Frange P, Cros G, Cavazzana M, André-Schmutz I, Schiff C (2014) The BLNK adaptor protein has a nonredundant role in human B-cell differentiation. J Allergy Clin Immunol 134(1):145–154
61. NaserEddin A, Shamriz O, Keller B, Alzyoud RM, Unger S, Fisch P, Prus E, Berkun Y, Averbuch D, Shaag A, Wahadneh AM, Conley ME, Warnatz K, Elpeleg O, Stepensky P (2015) Enteroviral infection in a patient with BLNK adaptor protein deficiency. J Clin Immunol 35(4):356–360
62. Conley ME, Dobbs AK, Quintana AM, Bosompem A, Wang YD, Coustan-Smith E, Smith AM, Perez EE, Murray PJ (2012) Agammaglobulinemia and absent B lineage cells in a patient lacking the p85ª subunit of PI3K. J Exp Med 209(3):463–470
63. Boisson B, Wang YD, Bosompem A, Ma CS, Lim A, Kochetkov T, Tangye SG, Casanova JL, Conley ME (2013) A recurrent dominant negative E47 mutation causes agammaglobulinemia and BCR- B cells. J Clin Invest 123(11):4781–4785
64. Borghesi L, Aites J, Nelson S, Lefterov P, James P, Gerstein R (2005) E47 is required for V(D)J recombinase activity in common lymphoid progenitors. J Exp Med 202(12):1669–1677
65. Kerns HM, Ryu BY, Stirling BV, Sather BD, Astrakhan A, Humblet-Baron S, Liggitt D, Rawlings DJ (2010) B-cell specific lentiviral gene therapy leads to sustained B cell recovery in a murine model of X-linked agammaglobulinemia. Blood 115(11):2146–2155
66. Bestas B, Turunen JJ, Blomberg KE, Wang Q, Månsson R, El Andaloussi S, Berglöf A, Smith CI (2015) Splice-correction strategies for treatment of X-linked Agammaglobulinemia. Curr Allergy Asthma Rep 15(3):510
67. Bachmeyer C, Monge M, Cazier A, Le Deist F, de Saint BG, Durandy A, Fischer A, Mougeot-Martin M (2000) Gastric adenocarcinoma in a patient with X-linked agammaglobulinemia. Eur J Gastroenterol Hepatol 12:1033–1035
68. Echave-Sustaeta JM, Villena V, Verdugo M, Lopez-Encuentra A, de Agustin P, Alberti N (2001) X-linked agammaglobulinemia and squamous lung cancer. Eur Respir J 17:570–572
69. Van de Meer JW, Weening RS, Schellekens PT, van Munster IP, Nagengast FM (1993) Colorectal cancer in patients with X-linked agammaglobulinemia. Lancet 341:1439–1440

Pulmonary Pathology in Agammaglobulinemia: Diagnosis and Treatment

3

Ulrich Baumann

3.1 Introduction and Clinical Phenotype

3.1.1 Respiratory Conditions Prior to Diagnosis

Deficiency of Bruton's tyrosine kinase and other genetic defects, which lead to maturation defects of B cells and are summarized as agammaglobulinemias, leads to clinical conditions primarily related to an increased susceptibility to infections at multiple organ sites. The airways are the most frequently affected site of infections with involvement of both the upper and the lower airways. This may not be a matter of surprise, for airway infections are also the leading cause to seek medical assistance in primary care [1]. However, airway infections in agammaglobulinemias differ from the usual common cold in that their course is more severe, protracted, and usually of bacterial origin.

The clinical picture caused by agammaglobulinemias is best characterized by the description of clinical features before the diagnosis is made and before the course of the disease is influenced by treatment with immunoglobulins. Data from various cohorts report affections of the respiratory tract prior to diagnosis in half to three quarters of the patients [2–7]. Some studies make a distinction between upper and of lower respiratory tract infections, with a frequency of lower airway infections between 29 and 65 % [2, 3, 6]. It is, however, likely that patients with agammaglobulinemia do suffer from upper as well as from lower respiratory tract infections. This is best described in an Italian cohort, of which only a minority was affected by upper airway infections alone, while more than half of the cohort suffered from both upper and lower airway infections [2]. Other reports do not make a distinction

U. Baumann
Paediatric Pulmonology, Allergy and Neonatology,
Hanover Medical School, Hannover, Germany
e-mail: Baumann.ulrich@mh-hannover.de

A. Plebani, V. Lougaris (eds.), *Agammaglobulinemia*,
Rare Diseases of the Immune System 4, DOI 10.1007/978-3-319-22714-6_3

between the sites of infections but focus on the incidence of pneumonia during the pre-diagnosis period. In these reports, the incidence of pneumonia ranged between 52 and 82 % [4, 5, 7], a frequency which is much in the same range as that of the reports on lower airway infections and thus perhaps addressing the same clinical condition. The fact that several authors prefer to report lower airway infections and not to pneumonias may, however, be due to the clinical picture of recurrent or chronic bacterial bronchitis which to our experience is a condition patients also frequently suffer from.

Upper airway infections appear to be conceived in the various cohort reports more broadly than by what is used in general practice. Two studies report on sinusitis [7] and, more specifically, chronic sinusitis [2] in the period prior to diagnosis at a sizeable proportion of their patients (37 and 21 %). These frequencies are uncommonly high compared to the healthy population and shed another light on the nature of airway disease in agammaglobulinemias. Although acute sinusitis may develop in the course of a common viral upper airway infection in healthy children in about every tenth episode [8], the diagnosis of sinusitis is usually not made before a more protracted course the infection evolves, because it is difficult to discriminate both conditions.

Acute bacterial sinusitis is considered in protracted upper airway infection episodes with nasal purulent discharge and obstruction persisting for more than 10 days [9]. Chronic sinusitis, as defined by nasal secretions and cough for more than 3 months, affected every fifth patient in the Italian cohort, before the diagnosis of the immunodeficiency was made [2]. This suggests that a sizeable proportion of patients reported to have suffered from sinusitis in fact had chronic sinusitis [7]. In the first years of life the size of the paranasal sinuses is minimal [10]. This may explain why chronic sinusitis is more likely to develop in school aged children in the general population and in patients with agammaglobulinemia [2, 11].

As probably the most common feature of lower respiratory tract infection, cough is surprisingly rarely reported in the literature. This does not necessarily mean that it is a negligible symptom. Given the definition of chronic sinusitis as used in the Italian cohort [2], it appears justified to assume that at least the same proportion of patients suffered from chronic cough as from chronic sinusitis. A high prevalence of chronic cough in undiagnosed patients with agammaglobulinemia is suggested also by the high proportion of patients who developed bronchiectasis and other structural lung disease already in the early phase of disease before the diagnosis of the immunodeficiency was made (see below paragraph on structural lung disease). Patients with structural changes of the bronchi are predisposed to prolonged episodes of productive cough due to an impaired mucociliary clearance (Fig. 3.1, page 18).

It is important to note that data on the clinical course and airway disease are almost exclusively available to patients with XLA. The diagnosis of XLA is usually straightforward on the basis of genetic testing, allowing for relatively homogeneous cohorts. Other variants of agammaglobulinemias are more rare and genetically diverse, or of unknown genetic etiology. The confinement on XLA cohorts does not mean that patients with autosomal recessive variants of agammaglobulinemia do considerably differ with their clinical presentation. It is more likely that they have a

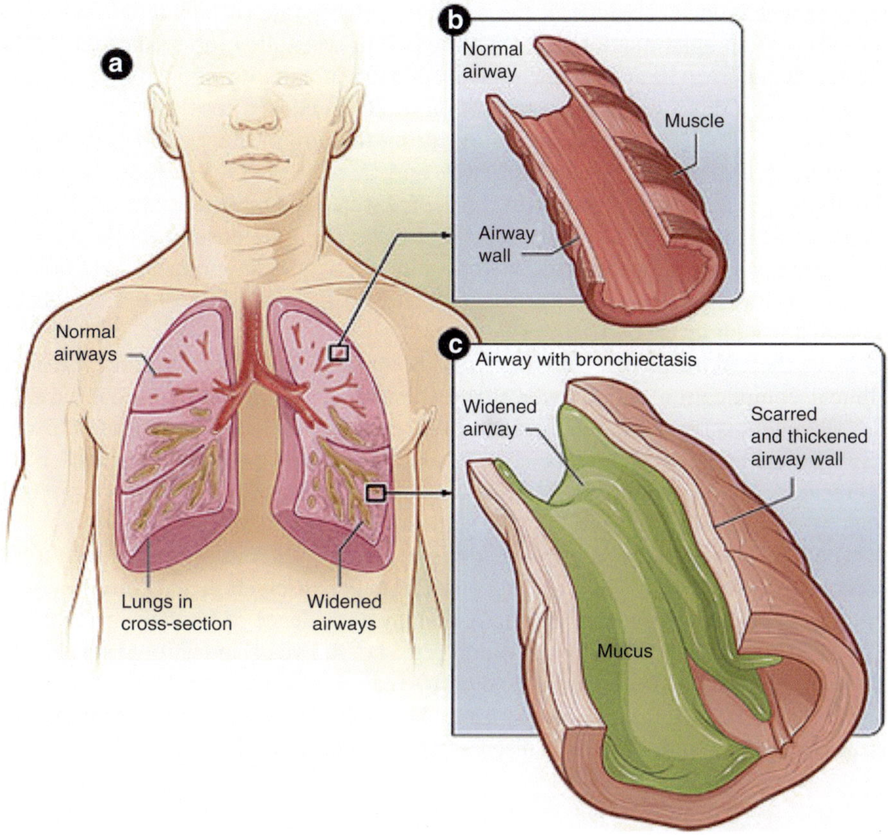

Fig. 3.1 Schematic view of bronchiectasis (**c**) indicating dilation and scarring of the bronchial wall, frequently accompanied by thickened airway wall and mucus obstruction of the airway lumen. (**a**) Shows a cross section of the lungs with normal airways and widened airways. (**b**) Shows a cross section of a normal airway (Source: http://en.wikipedia.org/wiki/File:Bronchiectasis_NHLBI.jpg, Wikipedia Commons)

clinical phenotype close to the XLA patients. Also patients with CVID show airway disease including sinusitis, pneumonia, and bronchiectasis at presentation at comparably high proportions [12].

3.1.2 Respiratory Conditions During Therapy

The introduction of immunoglobulin (Ig) replacement therapy and its later refinements have resulted in substantially improved survival rates and better health of the patients [2, 3, 7]. It seems therefore likely to assume that immunoglobulin replacement ameliorates the incidence and the course of respiratory conditions as the most frequent group of complications related to agammaglobulinemia. The effect of Ig replacement is most obvious in the reduction of the rate of pneumonia once the diagnosis of an agammaglobulinemia is made. In two follow-up studies on Italian XLA patients over 10 and 5 years, the proportion of patients who suffered from at least one episode of pneumonia had fallen to 26 % [2, 13], resulting in an annual incidence of 0.06 [2]. Pneumonia rates during therapy in other cohorts were also below the pre-diagnosis levels (Iran 38 % and USA 62 %, [4, 7]). Consistent with this finding is a meta-analysis of Ig replacement therapy (with approx. 40 % XLA patients) which showed a negative correlation between the dose or the IgG trough level and the rates of pneumonia, with an annual pneumonia incidence rate of 0.11 (95 % CI interval 0.04–0.3) at a trough level of 5 g/l and a rate of 0.023 (0.008–0.068) at trough levels of 10 g/l [14].

Nevertheless, respiratory tract infections in general remain the most frequent clinical complication under replacement therapy, although the empirical basis of this statement is less comprehensive [3]. A meta-analysis of studies with CVID and XLA patients receiving subcutaneous Ig therapy showed a negative correlation for IgG levels and the overall rate of infections, the majority of which were respiratory tract infections, only in a subgroup of studies [15].

The best evidence that upper and lower respiratory tract infections remain a not completely resolved clinical problem comes from the observation that chronic airway diseases appear to worsen also during Ig therapy. Prevalence of sinusitis increased by 22 % in a cohort over a mean observation period of 4.5 years and chronic sinusitis by 27 % during a mean follow-up period of 10 years [2, 7]. Concomitantly, the prevalence of chronic lung disease rose by 13 % during the same period [2]. The important issue of structural lung disease is discussed below in a separate chapter.

3.1.3 Structural Lung Disease

3.1.3.1 Bronchial and Parenchymal Disease

Chronic lung disease (CLD) is a term that was used to describe respiratory symptoms in conjunction with radiological pulmonary changes (bronchiectasis, bronchial wall thickening, or segmental atelectasis) in pioneering studies on lung disease with several large cohorts of immunodeficiency patients in Italy [2, 16]. CLD in our context must not be mixed with the condition of infants who require oxygen beyond the age of 28 days which is also termed CLD or bronchopulmonary dysplasia. Among the reported radiological changes, bronchiectasis is the key feature of CLD, by definition an irreversible dilation of a bronchus. Thickening of the bronchial wall can be a concomitant feature of bronchiectasis, indicating acute inflammation. Segmental atelectasis is the radiological feature of a collapsed segment of the lung, usually following obstruction of the related segmental bronchus.

At diagnosis, CLD was present in the Italian XLA cohort in 23 % of the patients. The probability of developing CLD during the period before diagnosis rose with time, for patients with CLD at presentation were on average older than those without [2]. After a mean observation time of 10 years and application of Ig therapy, however, prevalence of CLD had further risen to 36 %. The risk of developing CLD correlated with duration of therapy. The cumulative risk of developing CLD rose most rapidly in the interval between 10 and 25 years of follow-up from 0.07 to 0.92.

Other studies directly refer to the prevalence of bronchiectasis in cross-sectional cohorts, i.e., under Ig therapy. Bronchiectasis was reported at proportions of 23–26 % of the patients [4, 17]. Another Italian landmark study assessed prospectively the prevalence of bronchiectasis over a period of 5 years at which it rose from 33 to 39 % [13]. Focusing on radiological features of lung disease in more detail, a single-center study from Brescia (Italy) reported bronchial wall thickening and air trapping in only 1 of 12 patients with agammaglobulinemia, while 7 (58 %) had evidence of bronchiectasis. Parenchymal consolidation was present in 6 (50 %) and scaring in 4 (25 %) [18]. These findings suggest that CLD, as reported in [2], mainly represented bronchiectasis.

3.1.3.2 Functional Lung Disease

Lung function is a relatively easy to perform and inexpensive procedure to assess the physiology of the airways. However, lung function data of patients with agammaglobulinemia are sparse and limited to small cohorts. The high prevalence of bronchiectasis in agammaglobulinemia would suggest a significant proportion of obstructive lung disease which is not confirmed in the present data. In a cohort of 11 XLA patients, only 2 had mild reversible obstruction. Of note, spirometry in this report lead to normal findings, even if chest CT showed radiological changes, for 9 patients had structural lung pathology in CT [19]. In another study, at which CT was not performed on a routine basis, 9 out of 19 patients (47 %) had evidence of a restrictive pattern upon spirometry [4].

One study assessed spirometry longitudinally and calculated the annual decline of key parameters. Forced expiratory volume in 1 second (FEV_1) dropped by 65.6 ± 11 mL/year and forced vital capacity (FVC) 58.6 ± 20 mL/year. This decline was higher than in patients with CVID [20].

The available data is too limited to draw conclusions as to the nature of lung disease in agammaglobulinemia. There is, however, as yet no indication that spirometry is a sensitive tool for detection of bronchial pathology. It may serve better for longitudinal analysis based on calculation of the individual decline of FEV_1 or other spirometric volumes.

3.1.3.3 Mortality Related to Pulmonary Morbidity

Respiratory conditions, being the most frequent clinical complication prior to and during therapy, can take a devastating course and lead to lung transplantation or death. Second to disseminated enteroviral infection, cardiac failure due to chronic lung disease was the main cause of death reported in XLA cohorts [3, 7]. Lung transplantation also marks advanced stage pulmonary disease and is discussed in more detail below.

3.1.3.4 Infection

Although most of the pulmonary complications in agammaglobulinemia are considered to follow acute or chronic airway infection, data on pathogens are sparse. This may be due to the fact that there is no such pathogen of singular importance, as

Table 3.1 Pathogens and their frequency of detection in 20 out of 125 episodes of pneumonia in a cohort of XLA patients [7]

Pathogens and their frequency at pneumonia in XLA	
Streptococcus pneumonia	9
Haemophilus influenzae type B	5
Pseudomonas spp.	3
Staphylococcus aureus	3
Haemophilus parainfluenzae	3
Haemophilus parahaemolyticus	1
Klebsiella spp.	1
Mycobacterium avium	1
Pneumocystis carinii	1
Measles virus	1

P. aeruginosa in cystic fibrosis (CF), but rather a multitude of encapsulated cocci, gram positive and negative rods, other bacteria, and also viruses. Also, most infections are located at the lower airways and the paranasal sinuses. Specimens of these areas are more difficult to obtain and usually not part of routine assessment. Furthermore, it is likely that microorganisms we consider being pathogens are present along with commensal organisms in the areas more easily accessible, as the throat or the nasal turbinates. Further, pathogens may not come as a single infection, and it may be difficult for the laboratory to raise all different species growing at one plate. Finally, many patients already take antibiotics, when they come to medical attention which may prevent growth of relevant pathogens.

The best available data is from a cohort of XLA patients with documented 125 episodes of pneumonia, 20 of which had pathogens detected [7]. The species and their frequency is given in Table 3.1, page 18. It is important to note that two thirds of the pathogens were encapsulated. *H. influenzae and S. pneumoniae* were also the most frequently reported pathogens in other cohorts [6, 19]. However, it is equally of note that among the pathogens isolated in pneumonia there was a range of other pathogens that are primarily not associated with antibody deficiency, but with neutropenia (*P. aeruginosa*), T cell deficiency or phagocyte disorders (*Pneumocystis carinii*—today assigned as *P. jiroveci* and *M. avium*), or rarely pathogenic (*H. parainfluenzae and H. parahaemolyticus*) [7]. The notion that pathogens other than encapsulated bacteria are of particular relevance in XLA is further supported by the compilation of organisms isolated at episodes of septicemia in the same cohort, with *P. aeruginosa* as the most frequently isolated pathogen.

3.2 Diagnostic Methods

3.2.1 Chest X-Ray

Plain radiographs of the chest are available in standard settings of care, relatively cheap, and related to a very low irradiation dose. These advantages make chest X-ray a candidate for routine evaluation of lung disease in agammaglobulinemia, as

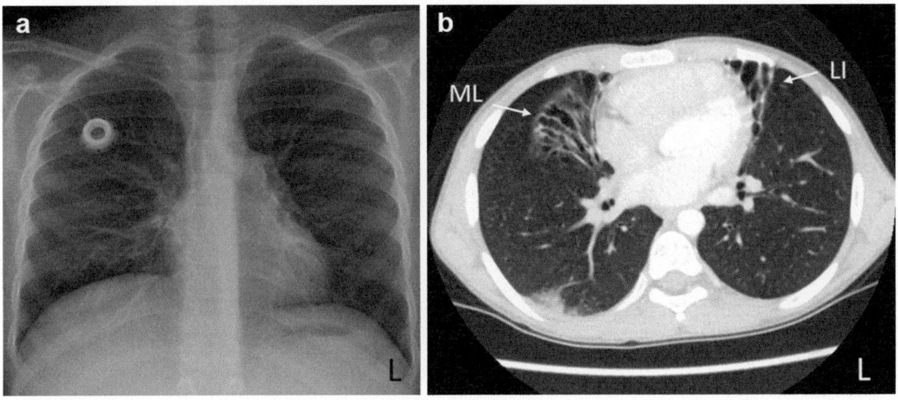

Fig. 3.2 Comparison of radiographs of the lung: plain X-ray (**a**) and transverse section of computed tomography (**b**). Bronchiectasis is evident in the middle lobe (*ML*) and the lingula (*LI*) with varicose and cystic bronchiectases cut longitudinally (*ML*) and transverse. Note that bronchiectasis is barely visible in plain chest radiograph (**a**) (Images kindly provided from Prof. Berthold. Paediatric Radiology, Hanover Medical School, Germany, EU)

it is common in other conditions with significant lung disease, such as CF [21]. Scores summarizing the degree of typical CF-related pathology, such as the Chrispin Norman score, allow to calculate the individual decline upon longitudinal assessment [22]. In agammaglobulinemia, however, chest X-ray has not reached a prominent position in routine care. It is recommended for baseline assessment for non-CF bronchiectasis and CVID patients in the UK to screen for consolidation and lung pathology in general, but not for routine follow-up [23, 24]. Chest X-ray is barely reported in the literature on agammaglobulinemia.

3.2.2 Chest CT

Computed tomography (CT) of the chest produces considerably more detailed information on pulmonary structures and is considered the gold standard for structural disease of the lung. CT allows for three-dimensional imaging, thus avoiding classical areas of "obscurity" of the lung, as the retrocardial parts of the lung. Bronchial structures can be assessed in detail. A meta-analysis of studies that report concomitant chest X-ray and CT findings in patients with XLA and CVID shows that chest X-ray fails to detect bronchiectasis in 31 out of 48 patients (65 %) [25–27]. An example on the poor representation of bronchiectasis in plain X-ray technique compared to chest CT is given in Fig. 3.2.

3.2.2.1 Bronchiectasis

Bronchiectasis is an irreversible dilation of a bronchus. In radiology, bronchiectasis refers to dilated bronchi only for bronchi with these changes are unlikely to regain their original shape. Dilation is measured by various ways (see Table 3.2, page 7,

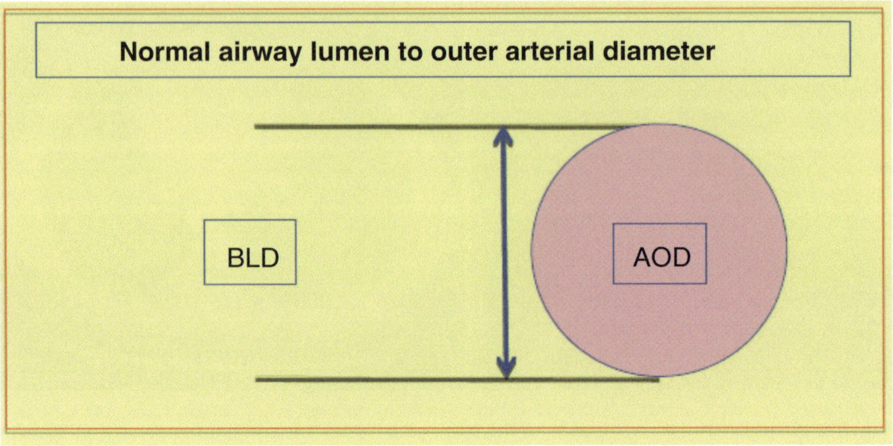

Fig. 3.3 Comparison of the diameter of the bronchial lumen (*BLD*) with the outer diameter of the accompanying artery (*AOD*). Bronchiectasis is defined as a BLD greater than the AOD (Graph taken from Frajoli [28])

Table 3.2 Radiological criteria of bronchiectasis in computed tomography of the chest

Signs of bronchiectasis in chest CT
Diameter of bronchial lumen > diameter of accompanying artery
Lack of tapering of a bronchus in its course to the periphery
Identification of bronchi within 1 cm of the pleural surface

Fig. 3.3, page 17, and images in [28]). According to its appearance, bronchiectasis can be described as cylindrical, varicose, or cystic which is, however, not elucidating the etiology. It is of more importance to quantify the degree of bronchiectasis. The size of a dilation of a given bronchiectasis can be measured and classified. An example of bronchiectasis grading is less than a third above the diameter of the accompanying artery, less than two thirds, or above [28]. It probably matters even more how much of the lung tissue is affected by bronchiectasis. This can be quantified in various ways, counting the number of lobes and/or estimating the proportion of the tissue that is affected. CT scores, such as the Brody score and its variations, use size and distribution of bronchiectasis a key features for calculation of severity grades [29, 30] (Table 3.2).

3.2.2.2 Bronchial Wall Thickening

CT severity scores also include other pathologies, such as bronchial wall thickening. In the healthy individual, the thickness of the bronchial wall is less than a third of the diameter of the accompanying artery (Fig. 3.4, page 21). Bronchial wall thickening can also be quantified by severity and distribution. Bronchial wall thickening is not necessarily associated with bronchiectasis. It is the radiological feature of edema and inflammation of the bronchial wall and may resolve without scarring or dilation. On the other hand, bronchiectasis may be the residual state of a dilated bronchus with no active inflammation, i.e., a bronchial wall with normal diameter, as seen in Fig. 3.4.

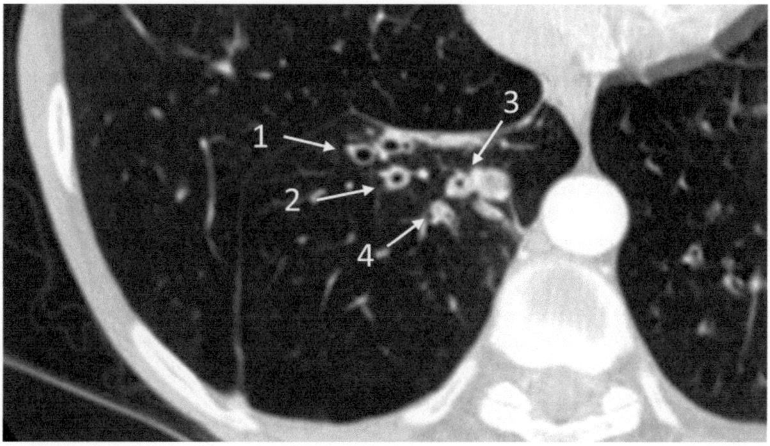

Fig. 3.4 Transverse section of computed tomography of the lung (detail of right lower lobe). *Arrows* indicate (*1*) mild bronchial wall thickening and bronchiectasis, (*2*) moderate bronchial wall thickening and bronchiectasis, (*3*) marked bronchial wall thickening, and (*4*) mucus plugging (Images kindly provided from Prof. Berthold. Paediatric Radiology, Hanover Medical School, Germany, EU)

3.2.2.3 Mucus Plugging

As the term describes, mucus plugging is an obstruction of the lumen of a bronchus without reference to pathology of the bronchial wall. The plug is an opaque filling of the bronchial lumen which normally is black due to the low radiodensity of air. If persisting, mucus plugging results in atelectasis of the downstream compartment of the lung. If a small airway is affected by mucus plugging, it is difficult to discriminate between the airway wall and its opacified lumen. This characteristic shape of mucus plugging in small bronchi is termed *tree-in-bud sign*.

3.2.2.4 Other Pulmonary Pathologies

Lung pathology as assessed by CT scan is not limited to the bronchi. Other changes that are considered to be located in the alveoli and in the interstitial tissue can be detected. Ground-glass appearance of various sizes and shapes indicates parenchymal or alveolar exudates. Bands and lines indicate fibrotic tissue, bullae a tissue defect, and emphysema a pathological ventilation. The full CT terminology [28, 31] is beyond the scope of this chapter, for it covers a wider range of pathologies than would be expected in agammaglobulinemia. In our own experience, however, lines, nodules, and emphysema affect between 10 and 20 % of the patients with XLA.

3.2.2.5 Present and Future Role of Chest CT

Chest CT is the most sensitive and accurate available technique to assess pulmonary pathology in agammaglobulinemia and antibody deficiencies. It depicts bronchial pathology long before spirometry or other lung function measures turn pathological. It allows quantifying pathologies which opens the perspective for interventional trials. Yet before chest CT finds a place in routine care, several major issues remain to be resolved.

CT produces images and not quantitative data. Assessment requires trained personnel and a considerable amount of time. Before CT data can be used meaningfully in multicentre studies, a common syllabus and a common catalogue which features have to be assessed and by which scales graded have to be agreed upon [31]. A European collaboration, the Chest CT in Antibody Deficiency Syndrome Group has focused on this issue and proposed a common list of CT findings relevant to patients with primary antibody deficiency in conjunction with a simplified score which is aimed to be applicable also in routine care (www.chest-ct-group.eu). Moreover, the Chest CT Group score is also compatible with one of the most frequently used scores for other conditions, a modified Brody score [29]. The Brody score with its variations is more detailed as it quantifies more in detail singular pathologies, making it more suitable to detect subtle progress of lung disease, as would be required in interventional studies. A follow-up study with CVID patients demonstrated that chest CT is sensitive enough to detect changes over a period of 5 years [32]. According to our own experience, chest CT scores can be sensitive enough for detecting even 2-year changes in patients with humoral immunodeficiencies.

Multicenter studies using chest CT require a common training of raters and quality control measures to guarantee inter- and intra-rater reliability, before this method can be successfully applied in interventional studies.

Use of chest CT in clinical practice varies considerably throughout Europe. Some national guidelines recommend chest CT as routine at regular intervals of 4 or 5 years. Others refer to chest CT only in case of pulmonary conditions which are difficult to manage. It is currently difficult to estimate the individual benefit of chest CT as long as there is no evidence which therapy is effective in conditions that are best detected by chest CT. For the time being, therefore, the decision of performing a chest CT should include the consideration whether the result would make a difference in care. A common documentation of any CT performed in a registry-like fashion would add a further benefit for the international community of immunodeficient patients. The Chest CT Group presently prepares this documentation.

Chest CT employs relatively high doses of irradiation compared to conventional chest films. Questions arising for patients with agammaglobulinemia can be dealt with at comparatively low doses for assessment of the bronchial structures and the ventilation state allows for a higher background noise and a lower number of sections than standard indications, like the search for metastases, or the assessment of interstitial lung disease. In most instances, use of contrast medium is dispensable. Observing these measures, the dose of a chest CT study can be substantially reduced [33]. Magnetic resonance imaging (MRI) bears no risk of radiation. It is too early to decide whether it will sufficiently substitute for the CT technique, but promising progress has been made also with patients with primary immunodeficiency [34].

3.2.3 Spirometry and Bodyplethysmography

Both methods measure static and dynamic volumes and flows. They have the unsurpassable advantage to be relatively easily accessible, cheap, and free of irradiation which makes them almost ideal for longitudinal management. FEV_1 is a key outcome parameter in numerous studies in other conditions that affect the lung, such as CF, non-CF bronchiectasis, or COPD. Yet the role of lung function in agammaglobulinemia remains to be determined. The available data show no consistent agammaglobulinemia-like pattern and course of lung function pathology. In comparison to chest CT, spirometry appears to be of low sensitivity to detect pulmonary pathology [19].

3.2.4 Induced Sputum

Induced sputum (IS) is a noninvasive method to achieve specimen from the bronchial airways by inhalation of hypertonic saline. It has the advantage that specimen can be obtained even from individuals who do not have spontaneous productive cough [35]. Assessment of IS allows for microbiological investigation as well as for cellular and soluble markers of inflammation. While initially research focused on the proportion of polymorphonuclear leukocytes as markers for inflammation, today cytokines such as interleukin 8 or tumor necrosis factor alpha are used as quantitative markers. The levels of these markers correlate positively with the severity of bronchiectasis in children with non-CF bronchiectasis [36]. IS was only occasionally used for research purposes in patients with agammaglobulinemia [37] and has not entered routine practice of care. It would, however, be a useful technique for microbiological assessment in patients aged above 5–6 years who do not produce sputum spontaneously and would allow a rational choice of antibiotic therapy.

3.3 Pathophysiology

Why are patients with agammaglobulinemia so extraordinarily susceptible to infections of the respiratory tract and less so at other organ sites, in particular when Ig replacement therapy is commenced? It is easy to conceive with respect to the immunological mechanisms that antibody deficiency results in susceptibility to pathogens that require antibodies to be removed from the body. This applies particularly to encapsulated bacteria which can effectively prevent their phagocytosis by a polysaccharide capsule which hides the pathogens under a relatively inconspicuous surface and by which phagocytes are normally not activated. This immune evasion allows the bacteria to colonize the mucosal surfaces of the body without causing an inflammatory reaction in a relatively peaceful coexistence as long as the bacteria do not cross the mucosal barrier. The high rates of pneumonia and invasive infection at multiple organ sites in agammaglobulinemic patients prior to therapy indicates that the bacteria do cross the mucosal barrier and that they are able to proliferate within the body. It is likely that crossing the barrier occurs in every individual at a low-level scale, but in health, the immune system is prepared to rapidly clear the pathogen without signs of disease.

However, it is less easy to conceive why pulmonary disease does not come to a halt by Ig replacement therapy, while invasive infections and even pneumonia are successfully prevented. All available evidence shows that bronchiectasis, sequelae of prolonged infection and inflammation of the bronchial mucosa, advances also during therapy [2, 13]. Several lines of evidence direct to potential causes of this disturbing conclusion.

3.3.1 Sinusitis as a Risk for Bronchial Infection

As bronchiectasis, sinusitis appears to advance during therapy. Some mild affection of the sinus mucosa may be present in almost all patients with agammaglobulinemia without overt sinusitis [18]. Sinusitis has a higher prevalence than bronchiectasis, and presence of sinusitis and bronchiectasis appears to correlate [13]. It is thus conceivable that bacteria which infect the nasal mucosa can be aspirated to the lower airways posing a small but repetitive risk for infection of the bronchi. Serial bronchoscopy along with microbiology of the nasopharynx shows in CF patients that *P. aeruginosa* infection always commences in the upper airways before it proceeds to the lower airways and not vice versa [38]. Molecular studies that compare the *P. aeruginosa* clones in the nasal and bronchial airways demonstrate a high concordance in the detection of pathogen clones in patients with cystic fibrosis, supporting the hypothesis that the bronchial clones derive from the upper airways [39]. The sinuses may also be a reservoir for reinfection in case of a successful clearance of the pathogens in the bronchi, for instance, by antibiotic therapy [40]. Colonization of the nasal airways can also explain the rapid infection with *P. aeruginosa* that follows implanted lungs from non-CF donors in CF lung transplant patients [41]. The nasal airways, in particular the paranasal sinuses, may serve also in patients with agammaglobulinemia as a site of primary infection, before the infection spreads to the lower airways.

3.3.2 Mucosal Antibody Deficiency

Antibody composition at the mucosal surfaces substantially differs to that of the systemic circulation. While IgG is the predominant isotype in plasma, polymeric IgA and IgM are the main isotypes at the surface of the respiratory mucosa [35, 42]. IgG reaches the airways by passive diffusion into the alveoli and moves to the central airways by mucociliary transport at decreasing concentrations. IgA and IgM reach the mucosal surface by active transepithelial transport by which IgA is also assembled to polymeric molecules with the secretory component. The fact that IgG therapy in patients with agammaglobulinemia is effective in preventing pneumonia, i.e., infection at the alveolar site, together with the observation that IgG therapy is less effective in preventing infections at the bronchial and paranasal sinus site, may be due to the relative low levels of IgG at the respiratory mucosa afar from the alveoli.

However, this hypothesis is not sufficient to satisfactorily explain the pulmonary phenotype in agammaglobulinemia. IgG levels at the respiratory mucosa might rise with the serum levels achieved by Ig replacement therapy and result in a better protection. Patients treated at higher IgG serum levels would have lower rates of bronchiectasis. Published data does not support this hypothesis well. Patients with bronchiectasis have similar or even higher serum IgG levels in XLA and CVID [13, 43]. Indeed, there may not even be a deficiency of IgG at the respiratory mucosal surface. Nasal washings of CVID patients show IgG at comparable or even higher levels than healthy controls [44]. A local inflammatory reaction, as shown in this study with elevated proinflammatory cytokines, may enhance local diffusion of systemic IgG to the mucosal surface also at the bronchial site. It is conceivable that IgG even augments the inflammatory response by its classical activation pathways on phagocytes and complement eventually leading to tissue damage by a long-term inflammatory response.

Other factors, therefore, may be more relevant for protection. In CVID patients, IgA deficiency is one of the few independent risk factors associated with bronchiectasis in multiple regression analysis [13]. CVID patients whose immunological phenotype includes intrinsic IgA production are better protected against chronic sinusitis, bronchiectasis, and pneumonia. It is likely that mucosal IgA is also present in these patients for primary antibody deficiency does not include deficiency of the secretory component. The role of IgA was frequently challenged by the observation that the most frequent humoral deficiency, selected IgA deficiency, commonly is not associated to a clinical phenotype [45]. Other factors might substitute for IgA deficiency at the mucosal surface, such as mucosal IgM. The role of IgM for mucosal protection, however, is less clear. Deficiency of systemic IgM in CVID and XLA is apparently less indicative for a pulmonary risk than IgA deficiency [13]. As to XLA, that most frequently includes IgA and IgM deficiency, it may be more a matter of surprise that not all patients are affected by bronchiectasis. But development of bronchiectasis may be only a matter of time, with a cumulative risk to developing CLD calculated as 92 % after 25 years of treatment [2].

It may be seen as discouraging that lung disease progresses during therapy in all XLA cohort studies that assessed this process. The view that we haven't done enough today by treating our patients with Ig replacement therapy can also stimulate our efforts, because much remains to be gained [46].

3.3.3 The Vicious Cycle of Infection and Inflammation

Different to CVID, patients with XLA and other variants of agammaglobulinemia have no indication for autoimmune inflammation of the lung in addition to their increased susceptibility of infections. The systemic and mucosal antibody deficiency in agammaglobulinemia is sufficient to explain the pattern of pulmonary infections with increased frequency, higher severity, and prolonged duration as characterized above. Bronchiectasis is not specific for antibody deficiency. The sequelae of an intense and/or prolonged inflammatory response at the bronchi are comparable at a number of conditions that are related to local infections with a pathological course [47]. The fact that bronchiectasis can also

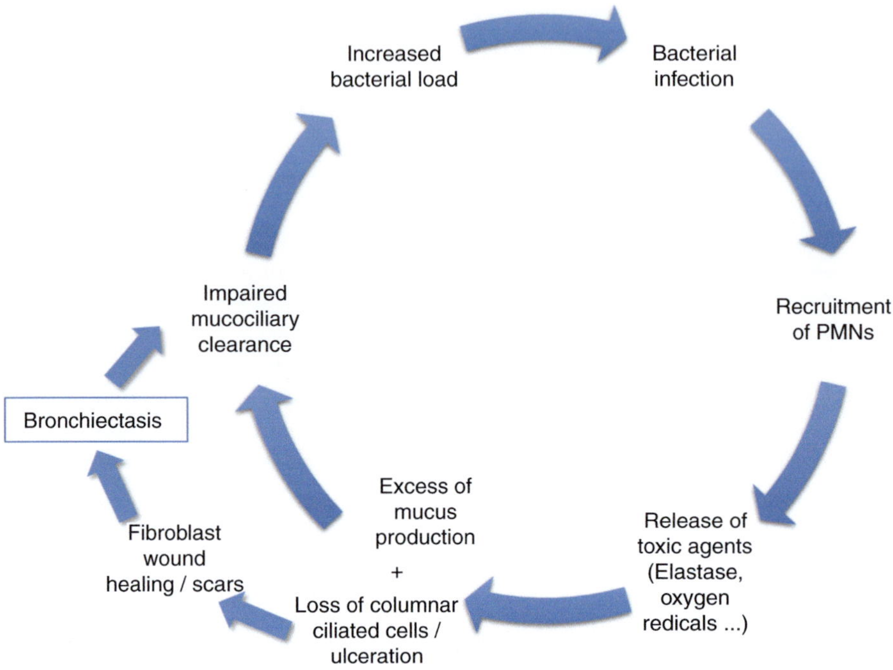

Fig. 3.5 Schematic diagram of bronchial infection and inflammation leading to bronchiectasis, termed as "vicious cycle" after [48]

evolve at conditions with no infectious etiology, such as idiopathic arthritis or celiac disease, directs to the understanding that the structural damage is caused by the inflammatory response and not be the pathogens directly. In agammaglobulinemia, as in every other condition, presence of bacteria at the mucosal surface activates resident macrophages and the respiratory epithelium to produce proinflammatory agents, such as interleukin 8 that recruits and activates neutrophils which migrate along the chemoattractant gradient to the site of infection. At the bronchial mucosa site, this universal mechanism of the immune response has several distinctive features. First, it is outside the body, and the supply of the cellular and humoral response is slower than at the tissue site. Second, inflammation leads to an excess of mucus production which can complicate the mucociliary clearance. Third, toxic agents, such as elastase or oxygen radicals, released by apoptotic neutrophils, can cause injury to the respiratory epithelium, resulting in a loss of the columnar ciliated cells and ulceration. This damage further compromises the mucociliary clearance which in turn prolongs the presence of bacteria and facilitates bacterial growth. If the epithelial injury is deep enough to result in airway remodelling and formation of scars, irreversible changes such as bronchiectasis are made. This concept, termed the vicious cycle of bronchiectasis (Fig. 3.5, page 23, [48]), is assumed for most conditions related to bronchiectasis [49].

This vicious cycle is an exception in the healthy individual and may occur following serious or unusual infections, such as pertussis, or atypical mycobacteria [50], while it occurs frequently in agammaglobulinemia.

The etiology as to why agammaglobulinemia has such a propensity to bronchiectasis may even extend beyond the deficiency in mucosal immunoglobulins. BTK is known to be expressed not only in B cells, but also in myeloid cells, attenuating the inflammatory response upon activation by several toll-like receptors (TLR) [51]. The response of BTK-deficient myeloid cells varies in vitro, showing no clear phenotype, but may be related to a more pronounced inflammatory response to bacterial antigens, such as LPS. The wide range of pathogens, patients with agammaglobulinemia are susceptible to despite Ig replacement therapy, would be compatible with a pathogenic role of BTK beyond B cell maturation.

3.4 Therapy

All patients with agammaglobulinemia are being treated with Ig replacement therapy. Its benefits and limitations are discussed in the above chapters. Here, we look at other therapies, their evidence, and their potential.

3.4.1 Oral Antibiotic Therapy

There is some limited evidence that antibiotic treatment is effective in patients with non-CF bronchiectasis. Short courses of antibiotics reduce sputum volume and purulence [52]. Studies on long-term antibiotic therapy gave conflicting evidence with two studies showing an overall improvement, but other studies could not confirm that long-term antibiotics either reduce the risk of exacerbations or improve lung function [53]. There is no study with patients with agammaglobulinemia.

Considering the frequency of bacterial growth in microbiologic evaluation of XLA patients, with the majority of specimen showing encapsulated bacteria, it appears however justified to use beta-lactam antibiotics as a first-line therapy in case of acute exacerbations. In contrast to standard care of otherwise healthy individuals, several authors suggest an early start of a treatment course of respiratory tract infections that are in healthy individuals most frequently of viral origin. Antibiotic treatment of agammaglobulinemic patients in these instances may serve to prevent bacterial superinfection that is more likely to occur once the integrity of the respiratory epithelium is compromised by the viral infection [54]. Early antibiotic therapy in patients with agammaglobulinemia may as well serve as a strategy to prevent proliferation of bacteria already prevalent at the airway surface. Since viral infections can be a frequent occurrence also in patients beyond early childhood, and since the patients are commonly aware of the early phase of respiratory tract infections, some guidelines suggest to providing patient-held antibiotics in order to facilitate an early start of therapy. On the same grounds of common sense, guidelines also suggest prolonged courses of 10 to 14 days [55]. Although the benefit of the prolonged treatment courses is not proven, it is not without consideration. The respiratory epithelium takes 2 to 3 weeks to recover its integrity after viral infections. During recovery, molecules, such as gangliosides, which serve bacteria for adhesion are exposed and increase the risk of a prolonged bacterial infection [54, 56].

Long-term antibiotic therapy appears less well supported by evidence and is not recommended in most guidelines. However, it may be considered in patients with chronic productive and purulent cough and the detection of bacteria sensitive to oral drugs. In our experience, the most frequent pathogen, *H. influenzae*, is rarely treated effectively with amoxicillin despite a fully sensitive pattern. It may be better treated with cefuroxime, or amoxicillin and clavulanic acid. Long-term treatment with antibiotics requires a careful monitoring of adverse effects. Patients need advice on the occurrence of enteritis with *C. difficile* and monitoring of hepatic function.

Following reports on substantial improvement by treatment with long-term low-dose macrolides of patients with diffuse panbronchiolitis, a condition that is related to the formation of a *P. aeruginosa* biofilm at the airway epithelium [57], azithromycin was proven effective to ameliorate the course of lung disease in cystic fibrosis [58] and, more recently, in patients with non-CF bronchiectasis. Although the latter patient group is likely to be heterogeneous, macrolide treatment resulted in reduction of frequency of exacerbations and improvement of spirometry values and in quality of life in both adults and children [59]. A reduction in frequency of exacerbations by maintenance macrolide therapy was also seen in patients with COPD [60]. However, there is currently no proof that macrolide therapy reverses more advanced stages of disease as shown by a lack of improvement of the 6-minute walk test distances or hospital admission rates in patients with non-CF bronchiectasis. Other conditions are currently being investigated for macrolide therapy, such as primary ciliary dyskinesia (PCD).

Numerous modes of action have been described by which macrolides exert their surprising efficacy in conditions which are to some extent resistant to conventional antibiotic therapy. Long-lived macrolides, such as azithromycin, achieve intracellular drug levels that reach the minimal inhibitory concentrations (MICs) of normally resistant pathogens, such as *P. aeruginosa*. Accordingly, drug concentrations several orders of magnitude above serum levels can be found in sputum of CF patients, for the sputum contains to a major proportion the cellular debris of phagocytes [61]. Moreover, sputum of CF patients treated with a 3 months course of azithromycin shows considerably improved rheological properties which is in the order of other mucolytic agents such as DNAse [61]. An important part of the effects of long-term low-dose macrolide therapy, however, applies to an attenuation or at least modulation of the host immune response to bacteria. Macrolides inhibit the synthesis of proinflammatory agents, such as interleukin 8, possibly by inhibition of NF-kappaB activation, and reduce the expression of adhesion molecules, such as ICAM-1. Subsequently, recruitment of neutrophils to the airway epithelium is reduced. Macrolides inhibit degranulation of neutrophils and reduce the activity of myeloperoxidase, nitric oxide, and malondialdehyde (summarized in [62]).

While macrolide therapy has proven successful in randomized controlled clinical trials with several conditions, important questions remain for use in standard practice. The maximum length of studies was 12 months. It is unclear whether the benefit of this treatment continues and whether adverse effects remain under control. Recently, an increased risk of acquisition of non-tuberculous mycobacteria for patients with CF receiving long-term macrolide therapy was discussed [63].

Attenuation of the inflammatory host response upon bacterial infection in patients with agammaglobulinemia may not necessarily be as beneficial as shown in other conditions which do not appear to have an intrinsic immunodeficiency but rather an augmented inflammatory response [64]. Use of long-term low-dose macrolides in agammaglobulinemia should, therefore, be reserved to randomized clinical trials.

3.4.2 Inhaled Antibiotic Therapy

Inhaled antibiotics, namely, tobramycin, colistin, aztreonam, and ciprofloxacin, are effective in reducing of rate of exacerbations and respiratory symptoms in patients with CF. They reduce sputum bacterial density of *P. aeruginosa* and increase the rate of eradication before a chronic airway infection is established [65, 66]. The efficacy of this approach is largely based on the impact on *P. aeruginosa* infection which explains the selection of the antibiotics used. Also in non-CF bronchiectasis, evidence is based on antipseudomonal effects of tobramycin, gentamicin, and colistin. Recently, a dry powder formulation of ciprofloxacin inhaled over a period of 4 weeks reduced the number of exacerbations and the total bacterial load in sputum. Interestingly, the reduced numbers of bacteria applied to *S. pneumoniae*, *H. influenzae,* and *M. catarrhalis*, but not to *S. aureus* [67]. Again, no studies are available on the use of inhaled antibiotics in patients with agammaglobulinemia. For *P. aeruginosa* is not a regularly detected pathogen in the airways of patients with agammaglobulinemia, it appears unlikely that the findings of trials in CF and non-CF bronchiectasis can be transferred to agammaglobulinemia easily. The differing range of pathogens, mainly encapsulated bacteria, such as *H. influenzae*, *S. aureus*, and *S. pneumoniae*, argues for the use of either fluoroquinolones or beta-lactam antibiotics. Experience with inhaled beta-lactam antibiotics is limited [68] and raises safety concerns on the development of allergic reactions. Thus, fluoroquinolones appear the first choice for trials in XLA and other conditions with hypo- or agammaglobulinemia. Due to the relatively low threshold in the development of antibiotic resistance, however, the issue of duration of therapy and of patient selection has to be resolved.

3.4.3 Inhaled Hyperosmolar Agents

Hypertonic saline and mannitol powder are thought to bind free water which subsequently improves the rheological properties of sputum present in the bronchial airways. This would facilitate sputum expectoration which in turn would reduce both bacterial load and the amount of host proteases. Bronchial damage by host proteases and oxygen radicals is considered the main pathophysiological mechanism that causes structural pathology of the bronchi, i.e., bronchiectasis [49].

Hypertonic saline is frequently used in CF. It is shown to reduce pulmonary exacerbations and improving the subjective feeling to have a clearer chest, but evidence for improvement of FEV_1 beyond the time of 4 weeks is weak [69]. With a small number of patients studied, there is as yet no proven benefit in patients with non-CF bronchiectasis [70].

In two phase III clinical trials, mannitol, a dry powder sugar alcohol, improved FEV_1 and reduced exacerbation rates in patients with cystic fibrosis over periods of up to 52 weeks [71, 72]. However, therapeutic strategies effective in CF are not necessarily effective in other chronic pulmonary conditions. Recently, mannitol inhalation failed to reduce the exacerbation rate in patients with non-CF bronchiectasis [73]. This trial does not exclude an effect of mannitol for the patients that suffered from a relatively mild disease, and it is possible that a therapeutic effect is more evident in patients with a higher burden of bronchiectasis and sputum production.

There are no studies on hyperosmolar agents in agammaglobulinemia patients. The relatively small costs and the good safety profile make both agents study candidates for XLA and ARA despite their failure in non-CF bronchiectasis patients.

3.4.4 Mucolytic Agents

DNA is a major contributor of sputum viscosity. The DNA derives from polymorphonuclear cells that succumbed after activation and release of proteases, in most instances following local bacterial infection. DNAse hydrolyses DNA polymers by shortening DNA fragments. Inhaled human recombinant DNAse (dornase alpha) is effective in reducing the frequency of exacerbations and improving spirometry in patients with CF under therapy for up to 2 years [74]. Since patients with non-CF bronchiectasis similarly suffer from recurrent or chronic bacterial infection and also have purulent sputum, it seemed likely that DNAse is beneficial in this patient group. However, transfer of this therapy was not straightforward. Several trials found no beneficial effect [74], and a larger trial even resulted in a greater decline of FEV_1 and a higher rate of exacerbation in the treatment group [75]. Thus, DNAse is commonly avoided in non-CF bronchiectasis.

Other mucolytic agents, bromhexine and erdosteine, were assessed for non-CF bronchiectasis in small trials only, demonstrating no consistent benefit [67].

There is no data on the effect of mucolytic agents in agammaglobulinemia. It is unclear whether DNAse would equally lead to adverse treatment outcome as in the non-CF bronchiectasis trial. DNAse may exert its action not only by a mucolytic effect, but may have also anti-inflammatory properties [76].

3.4.5 Airway Clearance Techniques

In CF, a broad range of techniques have evolved aiming to improve sputum expectoration. This comprises particular manners of breathing, coughing, postures, and percussion of the chest by a therapist (chest physiotherapy). Several tools can be used to produce an oscillating pulse wave of the air in the bronchi which is supposed to widen the airways and to open areas of bronchial collapse by a positive airway pressure. Oscillating devices are thought to improve the rheological properties of the sputum by exerting sheer forces on the molecules that keep the mucus tight with an oscillating change in pressure and airflow. A number of small studies have shown an increase in sputum expectoration. However, improvement of spirometric measures has occasionally been shown [77], but not as a long-term benefit [78]. In direct comparison, oscillating devices were as good as conventional chest physiotherapy in enhancing sputum expectoration [79].

Data on airway clearing techniques outside CF care are limited. Only few studies investigated the effect of various techniques in patients with non-CF bronchiectasis showing an increase in sputum volume or a decrease in functional residual capacity (FRC) by body plethysmography, considered as an improvement of hyperinflation [80].

While again there are no studies investigating airway clearance techniques with agammaglobulinemic patients, physicians in care of this patients frequently include chest physiotherapy into their therapeutic regimen and strongly recommended to do so in treatment recommendations [55]. This may be surprising at a time in which evidence-based recommendations are uniformly expected. The patients' experience of how rewarding it is to see a sizeable amount of sputum expectorated and how easy it can be to breathe after thorough chest physiotherapy perhaps explains better why a considerable number of patients and physicians favor this therapy.

3.4.6 Nasal Irrigation and Nasal Drug Deposition

The potential role of upper airway infections as a risk factor for lower airway infection is discussed above. It is conceivable that limiting the bacterial load of the nasal airways including the paranasal sinuses would reduce the risk of bacterial infection in the bronchial airways also in patients with agammaglobulinemia. Although none has been applied to agammaglobulinemia patients systematically, several therapeutic options could be beneficial for this purpose.

Nasal irrigation is a technique that originates from traditional Indian medicine and is used for treatment of acute and chronic sinusitis with some evidence of efficacy [81, 82]. The effect of nasal irrigation on bacterial infection at the lower airway is not investigated. Nasal irrigation is likely to reduce the amount of immunoglobulins on the epithelial surface. It remains to be seen whether the effect of bacterial (and viral) clearance outweighs the reduction in mucosal Ig concentration.

Topical deposition of drugs in the nasal airways is another approach aimed at reducing bacterial load and potentially relieving symptoms of sinusitis. A major obstacle, however, is that nasal irrigation, application of spray, or inhalation by conventional nebulizers bypass the paranasal sinuses. Nebulizers that superimpose an oscillating pulse wave upon the aerosol achieve high deposition rates in the paranasal cavities in vivo [83]. Nasal inhalation of tobramycin conveyed by a vibrant aerosol reduces the bacterial load of *P. aeruginosa* in patients with CF and improves clinical symptom scores of sinusitis [84]. Inhalation of DNAse with the same device also alleviates symptoms of sinusitis in CF. Moreover, these patients showed an improvement of vital capacity and of mean expiratory flow of mid-size airways (MEF_{75-25}) [85]. If this effect is not reached by bronchial inhalation of exceeding aerosol during nasal inhalation, treatment of the upper airways may indeed have indirect beneficial effects at the lower airways. There is still a long way to go until it becomes clear whether or not nasal pulse wave inhalation and paranasal drug delivery is beneficial in patients with agammaglobulinemia. Following the negative results of DNAse in non-CF bronchiectasis patients, other drugs would be preferable, in particular antibiotics that reduce the bacterial load of encapsulated bacteria at the nasal mucosal surface. Again, aminoglycosides or colistin do not appear ideal against the spectrum of pathogens predominant in the airways of patients with agammaglobulinemia.

3.4.7 Lung Transplantation

Only few reports exist on patients with XLA who underwent lung transplantation due to advanced stage lung disease [86, 87]. These patients exemplify the degree of lung disease that may be caused by the underlying disease. It is worth to look at these reports in more detail, for they not only report on end-stage lung disease that may develop in XLA and other agammaglobulinemias, they also make clear that lung transplantation is not straightforward in this group giving room for hypotheses as to why XLA patients fared worse than patients with other end-stage lung disease.

XLA patients who received lung transplant almost uniformly suffered from severe bronchiectasis in conjunction with infections by multiple species of encapsulated bacteria. Notably, no restrictive pattern, interstitial or granulomatous lung disease, as reported in CVID [88], or with bronchiolitis obliterans syndrome, a condition with a potential infectious etiology, were described. In a series of 6 Australian XLA patients, spirometry values and CO diffusion capacity were as low as in other patient groups considered for lung transplant (FEV_1 and DLCO 15 to 28 % and 34 to 58 % [pred.], respectively) [86]. All patients suffered from chronic productive cough and advanced stage bronchiectasis. In sputum cultures grew *H. influenzae* (3/6 patients), *Candida* spp. (2/6 patients), and *P. aeruginosa* (1 patient), presumably despite appropriate antimicrobial therapy.

The most remarkable detail of the Australian report is the ill-fated outcome of this small cohort. 5 out of 6 patients died within 4 years [86]. The other report on 2 patients from Spain had a follow-up for a maximum of 12 months which precludes conclusions as to the survival rate [87]. Four of the 5 deceased Australian patients died from intractable bacterial pneumonia and septicemia with a diverse spectrum of pathogens, including encapsulated gram-positive, gram-negative and intracellular bacteria, viruses, and fungi (*S. aureus*, *K. pneumoniae*, *S. pneumoniae*, *P. aeruginosa*, *M. avium*, *S. aureus*, *Giardia intestinalis*, *RSV*, *A. fumigatus*, *Zygomycetes*). Three patients also had evidence of bronchiolitis obliterans or graft rejection which did, however, not appear to be the leading cause of death, as it is common to other conditions after lung transplant.

The critical role of infections that complicated the post-transplant course while Ig replacement was continued at pre-transplant levels suggests that immunosuppression in BTK-deficient patients may increase the risk for uncontrolled infections more profoundly than in patients with other conditions, such as CF. The observation that most patients died from multiple infections including mycobacteria and fungi argues for a compromise in host defense mechanisms which is not necessarily restricted to Ig-dependent pathways. As discussed earlier, BTK is involved at the inflammatory response upon TLR stimulation, such as in formation of reactive oxygen species and other proinflammatory metabolites in myeloid cells of mice [89, 90]. The BTK defect in XLA may, therefore, affect the immune response beyond Ig formation and becoming clinically more relevant in the situation of immunosuppression. This would be consistent with the clinical observation of an increased susceptibility of immunosuppressed XLA patients to organisms which are normally cleared by phagocytes [7].

3.5 Outlook

The introduction of Ig replacement therapy substantially improved life expectancy and quality of life of patients with agammaglobulinemia. However, respiratory conditions, in particular chronic sinusitis and structural lung disease, continue to progress. New strategies for an improved treatment of airway disease are warranted. As long as corrective gene therapy is not available, topical antibiotic therapy, mucolytic agents, physiotherapy, and other measures that address the vicious cycle of infection and inflammation at the respiratory mucosa need to be studied for their potential to improve the life of the patients. Prerequisites of interventional trials are standardized and sensitive diagnostic procedures, such as chest CT, and microbiological and lung function monitoring, which is shared by a network of immunology centers. The Italian IPINET collaboration has demonstrated that much can be gained with this approach. Today, collaborations across Europe and beyond take shape in order to reach what still lies ahead of us: the challenging and rewarding goal to confine pulmonary complications and eventually to enable our patients to lead a healthy life.

References

1. Okkes IM, Polderman GO, Fryer GE, Yamada T, Bujak M, Oskam SK et al (2002) The role of family practice in different health care systems: a comparison of reasons for encounter, diagnoses, and interventions in primary care populations in the Netherlands, Japan, Poland, and the United States. J Fam Pract 51(1):72–73

2. Plebani A, Soresina A, Rondelli R, Amato GM, Azzari C, Cardinale F et al (2002) Clinical, immunological, and molecular analysis in a large cohort of patients with X-linked agammaglobulinemia: an Italian multicenter study. Clin Immunol 104(3):221–230
3. Lederman HM, Winkelstein JA (1985) X-linked agammaglobulinemia: an analysis of 96 patients. Medicine (Baltimore) 64(3):145–156
4. Aghamohammadi A, Allahverdi A, Abolhassani H, Moazzami K, Alizadeh H, Gharagozlou M et al (2010) Comparison of pulmonary diseases in common variable immunodeficiency and X-linked agammaglobulinaemia. Respirology 15(2):289–295
5. Conley ME, Howard V (2002) Clinical findings leading to the diagnosis of X-linked agammaglobulinemia. J Pediatr 141(4):566–571
6. Basile N, Danielian S, Oleastro M, Rosenzweig S, Prieto E, Rossi J et al (2009) Clinical and molecular analysis of 49 patients with X-linked agammaglobulinemia from a single center in Argentina. J Clin Immunol 29(1):123–129
7. Winkelstein JA, Marino MC, Lederman HM, Jones SM, Sullivan K, Burks AW et al (2006) X-linked agammaglobulinemia: report on a United States registry of 201 patients. Medicine (Baltimore) 85(4):193–202
8. Wald ER (1991) Purulent nasal discharge. Pediatr Infect Dis J 10(4):329–333
9. Meltzer EO, Hamilos DL, Hadley JA, Lanza DC, Marple BF, Nicklas RA et al (2006) Rhinosinusitis: developing guidance for clinical trials. J Allergy Clin Immunol 118(5 Suppl):S17–S61
10. Stenner M, Rudack C (2014) Diseases of the nose and paranasal sinuses in child. GMS Curr Top Otorhinolaryngol Head Neck Surg 13:Doc10
11. Stokken J, Gupta A, Krakovitz P, Anne S (2014) Rhinosinusitis in children: a comparison of patients requiring surgery for acute complications versus chronic disease. Am J Otolaryngol 35(5):641–646
12. Gathmann B, Mahlaoui N, CEREDIH, Gerard L, Oksenhendler E, Warnatz K, et al (2014) Clinical picture and treatment of 2212 patients with common variable immunodeficiency. J Allergy Clin Immunol 134(1):116–126
13. Quinti I, Soresina A, Guerra A, Rondelli R, Spadaro G, Agostini C et al (2011) Effectiveness of immunoglobulin replacement therapy on clinical outcome in patients with primary antibody deficiencies: results from a multicenter prospective cohort study. J Clin Immunol 31(3):315–322
14. Orange JS, Grossman WJ, Navickis RJ, Wilkes MM (2010) Impact of trough IgG on pneumonia incidence in primary immunodeficiency: a meta-analysis of clinical studies. Clin Immunol 137(1):21–30
15. Orange JS, Belohradsky BH, Berger M, Borte M, Hagan J, Jolles S et al (2012) Evaluation of correlation between dose and clinical outcomes in subcutaneous immunoglobulin replacement therapy. Clin Exp Immunol 169(2):172–181
16. Quinti I, Soresina A, Spadaro G, Martino S, Donnanno S, Agostini C et al (2007) Long-term follow-up and outcome of a large cohort of patients with common variable immunodeficiency. J Clin Immunol 27(3):308–316
17. Quartier P, Debre M, de Blic J, de Sauverzac R, Sayegh N, Jabado N et al (1999) Early and prolonged intravenous immunoglobulin replacement therapy in childhood agammaglobulinemia: a retrospective survey of 31 patients. J Pediatr 134(5):589–596
18. Bondioni MP, Duse M, Plebani A, Soresina A, Notarangelo LD, Berlucchi M et al (2007) Pulmonary and sinusal changes in 45 patients with primary immunodeficiencies: computed tomography evaluation. J Comput Assist Tomogr 31(4):620–628
19. Rusconi F, Panisi C, Dellepiane RM, Cardinale F, Chini L, Martire B et al (2003) Pulmonary and sinus diseases in primary humoral immunodeficiencies with chronic productive cough. Arch Dis Child 88(12):1101–1105
20. Chen Y, Stirling RG, Paul E, Hore-Lacy F, Thompson BR, Douglass JA (2011) Longitudinal decline in lung function in patients with primary immunoglobulin deficiencies. J Allergy Clin Immunol 127(6):1414–1417
21. Smyth AR, Bell SC, Bojcin S, Bryon M, Duff A, Flume P et al (2014) European cystic fibrosis society standards of care: best practice guidelines. J Cyst Fibros 13(Suppl 1):S23–S42

22. Benden C, Wallis C, Owens CM, Ridout DA, Dinwiddie R (2005) The Chrispin-Norman score in cystic fibrosis: doing away with the lateral view. Eur Respir J 26(5):894–897
23. Pasteur MC, Bilton D, Hill AT, British Thoracic Society Bronchiectasis Non-CF Guideline Group (2010) British Thoracic Society guideline for non-CF bronchiectasis. Thorax 65(Suppl 1):i1–i58
24. CVID diagnosis and management (Version No. 2.0) (2009). [Internet]. UK Primary Immunodeficiency Network, Newcastle upon Tyne [updated June 2009; cited 4th April 2015]. Available from: http://ukpin.org.uk/home/standards.htm
25. Kainulainen L, Varpula M, Liippo K, Svedstrom E, Nikoskelainen J, Ruuskanen O (1999) Pulmonary abnormalities in patients with primary hypogammaglobulinemia. J Allergy Clin Immunol 104(5):1031–1036
26. Newson T, Chippindale AJ, Cant AJ (1999) Computed tomography scan assessment of lung disease in primary immunodeficiencies. Eur J Pediatr 158(1):29–31
27. Thickett KM, Kumararatne DS, Banerjee AK, Dudley R, Stableforth DE (2002) Common variable immune deficiency: respiratory manifestations, pulmonary function and high-resolution CT scan findings. QJM 95(10):655–662
28. Frajoli F, for the Chest CT in Antibody Deficiency Group (2012) An image repository for chest CT [Internet]. EU, Hanover [updated July, 2012; cited 4th Apr 2015]. Available from: http://www.chest-ct-group.eu/downloads/findings/Chest%20CT%20Group%20Image%20Repository.pdf
29. Loeve M, Gerbrands K, Hop WC, Rosenfeld M, Hartmann IC, Tiddens HA (2011) Bronchiectasis and pulmonary exacerbations in children and young adults with cystic fibrosis. Chest 140(1):178–185
30. Brody AS, Sucharew H, Campbell JD, Millard SP, Molina PL, Klein JS et al (2005) Computed tomography correlates with pulmonary exacerbations in children with cystic fibrosis. Am J Respir Crit Care Med 172(9):1128–1132
31. Hansell DM, Bankier AA, MacMahon H, McLoud TC, Muller NL, Remy J (2008) Fleischner Society: glossary of terms for thoracic imaging. Radiology 246(3):697–722
32. Bondioni MP, Soresina A, Lougaris V, Gatta D, Plebani A, Maroldi R (2010) Common variable immunodeficiency: computed tomography evaluation of bronchopulmonary changes including nodular lesions in 40 patients. Correlation with clinical and immunological data. J Comput Assist Tomogr 34(3):395–401
33. Young C, Owens CM (2013) Pediatric computed tomography imaging guideline. Acta Radiol 54(9):998–1006
34. Serra G, Milito C, Mitrevski M, Granata G, Martini H, Pesce AM et al (2011) Lung MRI as a possible alternative to CT scan for patients with primary immune deficiencies and increased radiosensitivity. Chest 140(6):1581–1589
35. Baumann U, Gocke K, Gewecke B, Freihorst J, von Specht BU (2007) Assessment of pulmonary antibodies with induced sputum and bronchoalveolar lavage induced by nasal vaccination against Pseudomonas aeruginosa: a clinical phase I/II study. Respir Res 8:57
36. Guran T, Ersu R, Karadag B, Akpinar IN, Demirel GY, Hekim N et al (2007) Association between inflammatory markers in induced sputum and clinical characteristics in children with non-cystic fibrosis bronchiectasis. Pediatr Pulmonol 42(4):362–369
37. Kainulainen L, Vuorinen T, Rantakokko-Jalava K, Osterback R, Ruuskanen O (2010) Recurrent and persistent respiratory tract viral infections in patients with primary hypogammaglobulinemia. J Allergy Clin Immunol 126(1):120–126
38. Burns JL, Gibson RL, McNamara S, Yim D, Emerson J, Rosenfeld M et al (2001) Longitudinal assessment of Pseudomonas aeruginosa in young children with cystic fibrosis. J Infect Dis 183(3):444–452
39. Mainz JG, Naehrlich L, Schien M, Kading M, Schiller I, Mayr S et al (2009) Concordant genotype of upper and lower airways P aeruginosa and S aureus isolates in cystic fibrosis. Thorax 64(6):535–540
40. Berkhout MC, Rijntjes E, El Bouazzaoui LH, Fokkens WJ, Brimicombe RW, Heijerman HG (2013) Importance of bacteriology in upper airways of patients with Cystic Fibrosis. J Cyst Fibros 12(5):525–529

41. Vital D, Hofer M, Benden C, Holzmann D, Boehler A (2013) Impact of sinus surgery on pseudomonal airway colonization, bronchiolitis obliterans syndrome and survival in cystic fibrosis lung transplant recipients. Respiration 86(1):25–31
42. Pilette C, Ouadriri Y, Godding V, Vaerman JP, Sibille Y (2001) Lung mucosal immunity: immunoglobulin-A revisited. Eur Respir J 18(3):571–588, 20018590
43. Lucas M, Lee M, Lortan J, Lopez-Granados E, Misbah S, Chapel H (2010) Infection outcomes in patients with common variable immunodeficiency disorders: relationship to immunoglobulin therapy over 22 years. J Allergy Clin Immunol 125(6):1354, 1360.e4
44. Rose MA, Schubert R, Schmitt-Grohe S, Reichenbach J, Zielen S (2006) Immunoglobulins and inflammatory cytokines in nasal secretions in humoral immunodeficiencies. Laryngoscope 116(2):239–244
45. Yel L (2010) Selective IgA deficiency. J Clin Immunol 30(1):10–16
46. Baumann U, Miescher S, Vonarburg C (2014) Immunoglobulin replacement therapy in antibody deficiency syndromes: are we really doing enough? Clin Exp Immunol 178(Suppl 1):83–85
47. Goeminne P, Dupont L (2010) Non-cystic fibrosis bronchiectasis: diagnosis and management in 21st century. Postgrad Med J 86(1018):493–501
48. Cole PJ (1984) A new look at the pathogenesis, management of persistent bronchial sepsis: a 'vicious circle' hypothesis and its logical therapeutic connotations. In: Davies RJ (ed) Strategies for the management of chronic bronchial sepsis. Medicine Publishing Foundation, Oxford, pp 1–20
49. Fuschillo S, De Felice A, Balzano G (2008) Mucosal inflammation in idiopathic bronchiectasis: cellular and molecular mechanisms. Eur Respir J 31(2):396–406
50. Dhillon SS, Watanakunakorn C (2000) Lady Windermere syndrome: middle lobe bronchiectasis and Mycobacterium avium complex infection due to voluntary cough suppression. Clin Infect Dis 30(3):572–575
51. Marron TU, Yu JE, Cunningham-Rundles C (2012) Toll-like receptor function in primary B cell defects. Front Biosci (Elite Ed) 4:1853–1863
52. Chan TH, Ho SS, Lai CK, Cheung SW, Chan RC, Cheng AF et al (1996) Comparison of oral ciprofloxacin and amoxycillin in treating infective exacerbations of bronchiectasis in Hong Kong. Chemotherapy 42(2):150–156
53. Evans DJ, Bara AI, Greenstone M (2007) Prolonged antibiotics for purulent bronchiectasis in children and adults. Cochrane Database Syst Rev (2):CD001392
54. Kubesch P, Lingner M, Grothues D, Wehsling M, Tummler B (1988) Strategies of Pseudomonas aeruginosa to colonize and to persist in the cystic fibrosis lung. Scand J Gastroenterol Suppl 143:77–80, 19887372
55. Tarzi MD, Grigoriadou S, Carr SB, Kuitert LM, Longhurst HJ (2009) Clinical immunology review series: an approach to the management of pulmonary disease in primary antibody deficiency. Clin Exp Immunol 155(2):147–155
56. Baumann U, Fischer JJ, Gudowius P, Ligner M, Herrmann S, Tümmler B et al (2001) Buccal adherence of Pseudomonas aeruginosa in patients with cystic fibrosis under long-term therapy with azithromycin. Infection 29(1):7–11, 20018392
57. Kobayashi H, Takeda H, Sakayori S, Kawakami Y, Otsuka Y, Tamura M et al (1995) Study on azithromycin in treatment of diffuse panbronchiolitis. Kansenshogaku Zasshi 69(6):711–722
58. Saiman L, Marshall BC, Mayer-Hamblett N, Burns JL, Quittner AL, Cibene DA et al (2003) Azithromycin in patients with cystic fibrosis chronically infected with Pseudomonas aeruginosa: a randomized controlled trial. JAMA 290(13):1749–1756
59. Gao YH, Guan WJ, Xu G, Tang Y, Gao Y, Lin ZY et al (2014) Macrolide therapy in adults and children with non-cystic fibrosis bronchiectasis: a systematic review and meta-analysis. PLoS One 9(3), e90047
60. Albert RK, Connett J, Bailey WC, Casaburi R, Cooper JA Jr, Criner GJ et al (2011) Azithromycin for prevention of exacerbations of COPD. N Engl J Med 365(8):689–698
61. Baumann U, King M, App EM, Tai S, König A, Fischer JJ et al (2004) Long term azithromycin therapy in cystic fibrosis patients: a study on drug levels and sputum properties. Can Respir J 11(2):151–155

62. Zarogoulidis P, Papanas N, Kioumis I, Chatzaki E, Maltezos E, Zarogoulidis K (2012) Macrolides: from in vitro anti-inflammatory and immunomodulatory properties to clinical practice in respiratory diseases. Eur J Clin Pharmacol 68(5):479–503
63. Binder AM, Adjemian J, Olivier KN, Prevots DR (2013) Epidemiology of nontuberculous mycobacterial infections and associated chronic macrolide use among persons with cystic fibrosis. Am J Respir Crit Care Med 188(7):807–812
64. Grassme H, Jendrossek V, Riehle A, von Kurthy G, Berger J, Schwarz H et al (2003) Host defense against Pseudomonas aeruginosa requires ceramide-rich membrane rafts. Nat Med 9(3):322–330
65. Langton Hewer SC, Smyth AR (2014) Antibiotic strategies for eradicating Pseudomonas aeruginosa in people with cystic fibrosis. Cochrane Database Syst Rev 11:CD004197
66. Ryan G, Singh M, Dwan K (2011) Inhaled antibiotics for long-term therapy in cystic fibrosis. Cochrane Database Syst Rev (3):CD001021. doi(3):CD001021
67. Wilson R, Welte T, Polverino E, De Soyza A, Greville H, O'Donnell A et al (2013) Ciprofloxacin dry powder for inhalation in non-cystic fibrosis bronchiectasis: a phase II randomised study. Eur Respir J 41(5):1107–1115
68. Falagas ME, Trigkidis KK, Vardakas KZ (2015) Inhaled antibiotics beyond aminoglycosides, Polymyxins and aztreonam: a systematic review. Int J Antimicrob Agents 45(3):221–233
69. Wark P, McDonald VM (2009) Nebulised hypertonic saline for cystic fibrosis. Cochrane Database Syst Rev (2):CD001506. doi(2):CD001506
70. Hart A, Sugumar K, Milan SJ, Fowler SJ, Crossingham I (2014) Inhaled hyperosmolar agents for bronchiectasis. Cochrane Database Syst Rev 5:CD002996
71. Bilton D, Robinson P, Cooper P, Gallagher CG, Kolbe J, Fox H et al (2011) Inhaled dry powder mannitol in cystic fibrosis: an efficacy and safety study. Eur Respir J 38(5):1071–1080
72. Aitken ML, Bellon G, De Boeck K, Flume PA, Fox HG, Geller DE et al (2012) Long-term inhaled dry powder mannitol in cystic fibrosis: an international randomized study. Am J Respir Crit Care Med 185(6):645–652
73. Bilton D, Tino G, Barker AF, Chambers DC, De Soyza A, Dupont LJ et al (2014) Inhaled mannitol for non-cystic fibrosis bronchiectasis: a randomised, controlled trial. Thorax 69(12):1073–1079
74. Jones AP, Wallis C (2010) Dornase alfa for cystic fibrosis. Cochrane Database Syst Rev (3):CD001127. doi(3):CD001127
75. O'Donnell AE, Barker AF, Ilowite JS, Fick RB (1998) Treatment of idiopathic bronchiectasis with aerosolized recombinant human DNase I. rhDNase Study Group. Chest 113(5):1329–1334
76. Paul K, Rietschel E, Ballmann M, Griese M, Worlitzsch D, Shute J et al (2004) Effect of treatment with dornase alpha on airway inflammation in patients with cystic fibrosis. Am J Respir Crit Care Med 169(6):719–725
77. Pfleger A, Theissl B, Oberwaldner B, Zach MS (1992) Self-administered chest physiotherapy in cystic fibrosis: a comparative study of high-pressure PEP and autogenic drainage. Lung 170(6):323–330, 19924973
78. Warnock L, Gates A, van der Schans CP (2013) Chest physiotherapy compared to no chest physiotherapy for cystic fibrosis. Cochrane Database Syst Rev 9:CD001401
79. Morrison L, Agnew J (2014) Oscillating devices for airway clearance in people with cystic fibrosis. Cochrane Database Syst Rev 7:CD006842
80. Lee AL, Burge A, Holland AE (2013) Airway clearance techniques for bronchiectasis. Cochrane Database Syst Rev 5:CD008351
81. Adappa ND, Wei CC, Palmer JN (2012) Nasal irrigation with or without drugs: the evidence. Curr Opin Otolaryngol Head Neck Surg 20(1):53–57
82. Hermelingmeier KE, Weber RK, Hellmich M, Heubach CP, Mosges R (2012) Nasal irrigation as an adjunctive treatment in allergic rhinitis: a systematic review and meta-analysis. Am J Rhinol Allergy 26(5):e119–e125
83. Moller W, Schuschnig U, Khadem Saba G, Meyer G, Junge-Hulsing B, Keller M et al (2010) Pulsating aerosols for drug delivery to the sinuses in healthy volunteers. Otolaryngol Head Neck Surg 142(3):382–388

84. Mainz JG, Schadlich K, Schien C, Michl R, Schelhorn-Neise P, Koitschev A et al (2014) Sinonasal inhalation of tobramycin vibrating aerosol in cystic fibrosis patients with upper airway Pseudomonas aeruginosa colonization: results of a randomized, double-blind, placebo-controlled pilot study. Drug Des Devel Ther 8:209–217
85. Mainz JG, Schien C, Schiller I, Schadlich K, Koitschev A, Koitschev C et al (2014) Sinonasal inhalation of dornase alfa administered by vibrating aerosol to cystic fibrosis patients: a double-blind placebo-controlled cross-over trial. J Cyst Fibros 13(4):461–470
86. Barnes S, Kotecha S, Douglass JA, Paul E, Hore-Lacey F, Stirling R et al (2015) Evolving practice: X-linked agammaglobulinemia and lung transplantation. Am J Transplant 15(4): 1110–1113
87. Morales P, Hernandez D, Vicente R, Sole A, Moreno I, Torres JJ et al (2003) Lung transplantation in patients with x-linked agammaglobulinemia. Transplant Proc 35(5):1942–1943
88. Prasse A, Kayser G, Warnatz K (2013) Common variable immunodeficiency-associated granulomatous and interstitial lung disease. Curr Opin Pulm Med 19(5):503–509
89. Krupa A, Fol M, Rahman M, Stokes KY, Florence JM, Leskov IL et al (2014) Silencing Bruton's tyrosine kinase in alveolar neutrophils protects mice from LPS/immune complex-induced acute lung injury. Am J Physiol Lung Cell Mol Physiol 307(6):L435–L448
90. Mangla A, Khare A, Vineeth V, Panday NN, Mukhopadhyay A, Ravindran B et al (2004) Pleiotropic consequences of Bruton tyrosine kinase deficiency in myeloid lineages lead to poor inflammatory responses. Blood 104(4):1191–1197

4 Immunoglobulin Replacement Therapy: Past, Present, Future

Hans D. Ochs

4.1 Introduction

At the beginning of the twenty-first century, prescribers and users of serum immunoglobulin (Ig) derived from large pools of human plasma have a choice of multiple preparations that underwent testing for efficacy and safety, were shown to have an acceptable half-life, and to be well tolerated by antibody-deficient patients if given as recommended, either intravenously or subcutaneously, daily, weekly, or once a month. While there are still debates as to optimal dosing, mechanisms leading to adverse events and breakthrough infections in spite of IgG trough levels that exceed recommended values, there is consensus that Ig replacement therapy improves quality of life and saves lives. To arrive at this state of success, the concepts of immunity, protective antibodies, and serum therapy had to be developed. The introduction of rigorous experimentation in the second half of the nineteenth century led to the discovery of pathogenic microbes and provided proof that infectious diseases are caused by different types of microoganisms. The concept of germs causing disease, developed in the late nineteenth century, was rationalized by Robert Koch: "On many occasions, I examined normal blood and normal tissue and there was no possibility of overlooking bacteria. Thus, I conclude that bacteria do not occur in healthy human or animal tissue." Louis Pasteur, by introducing the concept of biogenesis (*Omne vivum ex vivo*, "all life from life"), refuted the century-old belief in "spontaneous generation" and set the stage for using attenuated microbes such as anthrax (1881) and rabies (1885) to induce active immunity [1].

H.D. Ochs
Department of Pediatrics, University of Washington School of Medicine
and Seattle Children's Research Institute, Seattle, WA, USA
e-mail: allgau@uw.edu

A. Plebani, V. Lougaris (eds.), *Agammaglobulinemia*,
Rare Diseases of the Immune System 4, DOI 10.1007/978-3-319-22714-6_4

Emil von Behring, who in 1889 joined Robert Koch's Institute in Berlin, developed, together with Shibasaburo Kitasato, the experimental basis of serum therapy and of active or passive immunization [2, 3]. Using rabbits and mice, Behring and Kitasato discovered that blood and serum from animals that were rendered immune to tetanus neutralized or destroyed tetanus toxin. This property of serum was stable and could be transferred to other animals, which became protected from the toxic effects for a limited time. For this discovery, Behring received the first Nobel Prize in Medicine in 1901. This activity, observed only in sera from immunized animals, was heat stable (60 °C), in contrast to the bactericidal substance present in fresh blood from non-immunized animals, first recognized in 1889 by Buchner and subsequently named complement, which was heat labile [4]. Based on Behring's seminal experiments, Paul Ehrlich developed his "side chain" theory [5, 6]. Ehrlich proposed the presence of "antigen" receptors (side chains) on the surface of cells, which upon stimulation by antigen produce and secrete large quantities of "side chains" (antibodies) that bind to and mop up toxins. Thus, Ehrlich predicted the present concept of antigen-receptor-bearing B cells that upon activation produce high affinity antibody, an idea that was honored in 1908 with the Nobel Prize in Medicine. Behring and Ehrlich subsequently collaborated in the production and standardization of diphtheria-anti-toxin containing serum generated in cows, and created a company in Marbug now known as Behring Werke. The standardized serum was sold successfully in Europe and the USA. Serum therapy, as it was known, became widely used as anti-infectious and anti-toxin therapy [7]. Because of serious complications such as serum sickness and anaphylaxis, therapeutic sera generated in different animal species were required. It was soon discovered—as early as 1896—that these problems could be avoided by using convalescent human serum for the prevention of measles in exposed children [8, 9], by producing extracts from human placenta [10] or by isolating the globulin fraction of immune serum [11]. The separation and systematic analysis of serum proteins were made possible by the discovery in 1938 of serum electrophoresis by Tiselius and Cabat, proving that antibody activity was present in the gammaglobulin fraction of serum [12]. In response to the request of the National Research Council on behalf of the US Army, biochemists Edwin Cohn and John Oncley designed a method by which all major components of human plasma could be isolated as more or less homogeneous fractions, while the different components, including those known to be labile, were functionally preserved [13]. By using different concentrations of ammonium sulfate, increasing percentages of cold ethanol and by manipulating the pH, Cohn and Oncley obtained six fractions, with fraction II containing the gammaglobulin/antibody portion [14]. In a single volume of the Journal of Clinical Investigation (Volume 23, February 17, 1944) a series of reports were published that characterized the six protein fractions [15], and demonstrated that the gammaglobulin containing fraction II includes specific antibodies [14], and that concentrated gammaglobulin derived from normal adult serum was highly effective in preventing or attenuating measles when given within a week of exposure [16, 17]. These studies were extended to treat a long list of viral diseases with fraction II, including hepatitis A, respiratory syncytial virus, parvovirus B 19, and enteroviruses (summarized by Keller and Stiehm, 2000) [18]. When Ogden Bruton discovered agammaglobulinemia in 1952, he elected to try this new product in his 8-year-old patient who had suffered from multiple episodes of invasive pneumococcal disease, and demonstrated that subcutaneous injection of Cohn fraction II at a dose of 100 mg per kilogram per month, as suggested by Janeway, completely prevented the recurrent infections [19].

4.2 Early Use of Gammaglobulin as Replacement Therapy for PID

Following Bruton's report, similar patients with agamma-/hypogammaglobulinemia were identified [20, 21] and were treated with Cohn fraction II, available as a 16 % gammaglobulin solution in glycine buffer. While Bruton and Janeway preferred subcutaneous injection, in subsequent years, the material was given intramuscularly until intravenous formulations became available in the 1980s. Although intramuscular injections were effective, the shots were painful, the material was slowly and incompletely absorbed, and the quantity that could be injected at one time was severely limited.

Attempts were made to give unmodified gammaglobulin intravenously, but when infused into acutely ill children or hypogammaglobulinemic patients, severe vasomotor reactions or anaphylactic shock were frequently observed. Such reactions occurred occasionally even in healthy volunteers, receiving as little as 2 ml of unmodified fraction II intravenously [22]. To better understand the mechanisms that caused these reactions, Silvio Barandun and colleagues performed a series of daring in vivo experiments [23]. Working at the Tiefenau Hospital in Bern, Switzerland, they injected up to 10 ml of diluted standard IgG preparation (1.45 % gammaglobulin) into normal controls and patients with the diagnosis of antibody deficiency. Of the 55 normal controls receiving this material, seven (12.7 %) developed symptoms of flushing, chills, fever, and lumbar pain, while 14 of 15 patients with antibody deficiency (93 %) developed adverse events, including anaphylactic shock. A substantial decline in complement was detected in those individuals who developed symptoms. Interestingly, these investigators noticed that when the symptoms, caused by the initial infusion, had disappeared, the patients (and controls) became "refractory" to subsequent gammaglobulin infusions for up to 5 days [23].

Possible explanations of these anaphylactoid reactions considered by Barandun and colleagues included direct activation of complement by aggregated IgG (which in Cohn fraction II is estimated to be 5–10 % of the total IgG) or the formation of complexes between the infused IgG and antigens that, because of the lack of antibody, had accumulated in the patients' circulation, causing a "phlogistic reaction" [24]. Based on these clinical observations, the group in Bern explored innovative formulations that would be suitable for intravenous use. One strategy was to remove IgG aggregates using ultra centrifugation or nanofiltration; another was exposure to low pH and treatment with pepsin at low concentration insufficient to cleave peptide bonds in the native monomeric IgG molecule but sufficient to degrade the IgG aggregates. The resulting preparation, Sandoglobulin®, underwent nanofiltration, was exposed to pH 4, contained pepsin (1:10,000) and sucrose (1.676 g/1 g protein), was lyophilized, and could be reconstituted as 3–12 % IgG concentrations with osmolalities of 192–1074 mOsmol/L. Extensive clinical studies demonstrated excellent safety and efficacy as replacement therapy for patients with antibody deficiency [24]. In the USA, Cutter Laboratories decided to find a marketable use for the material contained in Cohn fraction II (other than selling it as fertilizer), a by-product of albumin fractionation from pooled human serum. Realizing that the demand for 16 % gammaglobulin that had to be injected intramuscularly was limited, Cutter approached the problem of aggregated IgG by employing reduction (dithiothreitol) and alkylation (iodoacetamide) to get rid of the aggregates. In a controlled crossover study, 34 patients with antibody deficiency were treated for 1 year with either 16 % immune serum globulin given intramuscularly or 5 % modified (reduced and alkylated) IVIG, both at a monthly dose of 100 mg/Kg. Disappointingly, those patients receiving Cutter-IVIG had a slight but statistically significant increase in the incidence of acute upper respiratory infections and twice as many episodes of adverse events [25]. During this clinical trial, two relevant observations were made: (1) If the lyophilized material was diluted in 5 % glucose instead of normal saline, the solute became clear and the infusions were tolerated better. (2) A small dose of the material (1–2 ml), given intravenously the day before the monthly infusions, prevented most of the clinical side effects such as flushing, headache, fever, and muscle ache. Based on the results of the 2-year crossover study, a follow-up double-blind crossover trial was designed to compare the 5 % reduced and alkylated formulation used previously with one to which 10 % maltose had been added. Each patient received three monthly IgG infusions of 100–250 mg/kg before being switched to the alternate preparation. Only 3 of 29 patients had adverse reactions when on the maltose-containing IVIG, compared with 22 of 29 having adverse events when receiving IVIG without maltose [26]. In vitro studies supported these clinical observations: fraction II, incubated for 24 h at 61 °C, formed extensive aggregates with increased optical density (OD = 0.92); addition of 10 % maltose prevented aggregate formation during exposure to 61 °C and showed a similar low OD (0.05) as the non-heated material, while the Cutter preparation containing 10 % maltose maintained a low OD (0.02) under the same conditions (Fig. 4.1). As a result, the maltose-containing Cutter product became the first FDA-approved IVIG preparation in the USA.

These clinical trials demonstrated that it was possible to modify Cohn fraction II and obtain formulations that could be given safely in large doses intravenously, opening a new era of immunoglobulin therapy, not only as a substitution for patients with antibody deficiency, but also immunomodulatory treatment for those with autoimmune disorders. Recognizing the importance of IVIG, the Bureau of Biologics (FDA), in conjunction with the National Heart, Lung and Blood Institute (NIH) sponsored an immunoglobulin workshop which took place during October 1979 in Bethesda, MD. At this consensus meeting, an international group of biochemists, immunologists, and clinicians as well as representatives from industry discussed the role and benefit of IVIG in the treatment of antibody-deficient patients, other potential indications, the prevention of adverse reactions, and methods to improve the safety of immunoglobulin products [27].

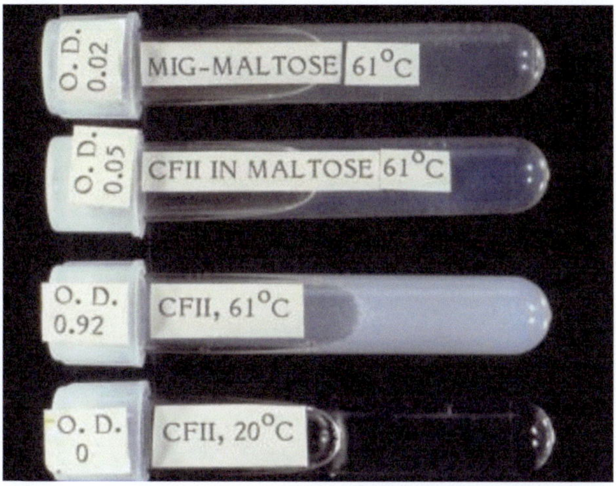

Fig. 4.1 Effect of sugar (10 % maltose) on heat-induced aggregation of IgG. Cohn fraction II (CFII) at room temperature (20 °C) remains clear with an optical density (OD) of 0 (bottom tube); if incubated for 24 h at 61° C, the solution turns cloudy with an OD of 0.92. This heat-induced aggregate of CFII is almost completely prevented when 10 % maltose is added (OD 0.05). The 5 % modified IgG (MIG) formulation (reduced and alkylated in 10 % maltose, developed by Cutter Laboratories), when exposed to 61° C for 24 h, does not form aggregates (OD 0.02)

In subsequent years, the goals established at the Bethesda meeting were reached and a number of highly effective, standardized, well-tolerated preparations are now available (Table 4.1). All FDA-approved formulations use highly purified biologically active (native) IgG, which are free of aggregates (<1 %), preservatives, and infectious agents. These preparations are produced from large donor pools, guaranteeing a broad spectrum of antibody activity and have normal distribution of IgG subclasses. With the exception of two products that are lyphoized, the preparations on the US market are available as 5 %, 10, or 20 % solutions; some preparations have added sugars for stabilization, others do not; most are slightly acidic with a pH of 4.0–6.0. Depending on the sugar content, the osmolality may reach unphysiologic levels.

To be licensed, each preparation had to be vigorously tested in clinical trials for safety, half-life, and therapeutic benefits.

Several incidences of transmission of hepatitis C occurred, either in lots of not-yet-licensed preparations that were in clinical trials [28] or in approved IVIG products [29, 30], but none occurred after 1994. No other pathogen transmissions such as HIV or prions have been reported. This excellent safety record has been attributed to the fractionation process, solvent/detergent (S/D) treatment, low pH (pH4) exposure, addition of octanoid acid (caprylate), pasteurization, nanofiltration, stringent donor screening, quarantine (hold back) programs, and the use of PCR-based testing.

In spite of these efforts to generate IVIG preparations with improved tolerability, episodes of adverse events (AE) still occur, including serious AEs [31]. Immediate

Table 4.1 Characteristics of immunoglobulin preparations currently available in the US

Brand name, manufacturer	Physical state, concentration	Shelf life at temperature	Stabilizer	NaCl content (mg/ml)	Osmolality (mOsm/kg)	IgA content (mcg/ml)	Approved administration	pH
Gammagard S/D, Baxter	Lyophilized 5 %, 10 %	24 months, room temperature (RT)	Glucose	8.5	636 (5 %) 1250 (10 %)	<2.2 <4.5	IV	6.4–7.2
Gammagard liquid, Baxter	Liquid 10 %	36 months (2–4 °C) 6 months, RT	Glycine	None	240–300	<37	IV SC	4.6–5.1
Gammaplex, Bioproducts Lab	Liquid 5 %	24 months, RT	D-Sorbitol 5 %	2.3	450–500	<10	IV	4.6–5.1
Carimune NF, CSL Behring	Lyophilized 3–12 %	24 months, RT	Sucrose (1.67 g/1 g protein)	<20 mg per g protein	192–1074	<720	IV	6.4–6.8
Privigen, CSL Behring	Liquid 10 %	36 months, RT	L-Proline, 250 mmol/L	<5	320	<25	IV	4.6–5.0
Hizentra, CSL Behring	Liquid 20 %	18 months, RT	L-Proline, 210 mmol/L, polysorbate 30 mg/l	Trace	280	≤50	SC (only)	4.6–5.2
Gamunex, Grifols (same as Gammaked, Kedrion)	Liquid 10 %	36 months, RT	Glycine	Trace	258	46	IV SC	4.0–4.5

Flebogamma 5 % Dif, Grifols	Liquid 5 %	24 months, RT	D-Sorbitol, 5 %	<3.2	240–370	<50	IV	5–6
Flebogamma 10 % Dif, Grifols	Liquid 10 %	24 months, RT	D-Sorbitol, 5 %	Trace	240–370	<100	IV	5–6
Octagam 5 %, Octapharma	Liquid 5 %	24 months, RT	Maltose 100 mg/ml	<3	310–380	<100	IV	5.1–6
Octagam 10 %, Octapharma	Liquid 10 %	24 months (2–8 °C) 6 months, RT	Maltose 90 mg/ml	<1	310–380	106	IV	4.5–5.5
Gammanorm, Octapharma	Liquid 16 %	36 months (2–8 °C)	Glycine polysorbate	2.5		83	IM SC	6.4–7.2
HyQvia, Baxter	Dual vial unit (1) 10 % IgG, same properties as Gammagard liquid (2) Recombinant human hyaluronidase (160 u/ml) (800 u/10 g) IgG				240–300	<37	SC (only)	4.6–5.1

infusion-related reactions are usually mild to moderate, occurring in 5–15 % of infusions overall and are experienced by approximately 20 % of patients over time [32]. These reactions, consisting of fever, chills, flushing, malaise, myalgia, flu-like symptoms, nausea, vomiting, and/or diarrhea, are experienced by up to 50 % of adult CVID patients receiving their first IVIG infusion. Therefore, the first and second IVIG infusions should be given in an infusion center by experienced staff. Infusion-related AEs can be prevented/treated by slowing or temporarily stopping the infusion, or by using anti-inflammatory drugs (Tylenol, ibuprofen, or hydrocortisone (50–100 mg)). Increasing the dose or reducing the interval between infusions (e.g., from every 4 to every 3 weeks) often eliminates these unpleasant AEs.

More serious anaphylactoid reactions resulting in bronchospasms, wheezing, and changes in vital signs are rare and may require stopping the infusion or initiating pretreatment with antihistamines or steroids. The role of IgG/IgE anti-IgA antibodies in those reactions is controversial and of minor significance.

The tendency to infuse large volumes of IVIG at a rapid rate has created "organ-specific" AEs. The most common reactions include "aseptic meningitis" syndrome beginning up to 72 h post-infusion and may be associated with stiff neck and spinal fluid leukocytosis [31, 33]. Acute renal failure has been observed, often in older patients receiving high doses of an IVIG preparation that contains sugars, mainly sucrose or maltose, as stabilizers, increasing the osmolality to unphysiologic levels. Preexisting renal disease is a frequent facilitator [34]. Other risk factors include diabetes, vascular disease, and dysproteinemia such as multiple myeloma or cryoglobulinemia [31]. Thrombotic events are predominantly arterial, causing myocardial infarction within hours or a few days following IVIG infusions. Venous thrombosis generally occurs days to weeks post-infusion and may affect deep veins, causing pulmonary embolism, CNS thrombosis, or jugular thrombosis [31]. Suspected mechanisms include indwelling venous catheters (e.g., Portacath), hyperviscosity, hypercoagulation due to autoantibody, or the presence of contaminating coagulation factors not removed during the fractionation process. In 2010, a sudden increase in thromboembolic events occurred following infusion of Octagam 5 %, resulting in the recall of the product. A change in the fractionation process had allowed the passage of activated FXI into the final product, which was identified as the major biological root cause of the thromboembolic events [35]. The presence of anti-A or anti-B IgG-isohemagglutinins may cause a positive direct Coombs test but rarely results in hemolysis [31, 36]; in rare instances, significant hemolysis can occur [37].

4.3 IG Replacement Therapy

The goal of Ig replacement is to ensure adequate passive immunity by providing physiologic levels of antibodies to protect deficient patients from common serious infections. It is recommended that the plasma pool collected for the production of IVIG consist of at least 1000 individual donors, preferably from the region where the product is intended for use. The final product must contain a broad spectrum of specific antibodies with titers to selected microorganisms, such as measles, that conform with the FDA requirement; The IgG has to be biologically active (native), with a physiologic half-life of 20–30 days, free of aggregates, and should represent all IgG subclasses proportionally. Unfortunately, antibody titers to commonly encountered microorganisms may vary not only between preparations, but also from lot to lot. This variability makes a treatment regimen a guessing game and supports the idea of individualizing immunoglobulin dose and delivery in order to optimize outcomes and costs [38, 39]. When first approved for clinical use in the 1980s, the recommended replacement dose for IVIG was 100–200 mg/kg per month, not enough to increase the trough level much above the base line (Fig. 4.2). In subsequent years, the dose requirement was increased to 400–600 mg/kg or higher, resulting in substantially higher trough levels (Fig. 4.2) and improved outcomes [40–42]. A meta-analysis of 17 clinical trials provided evidence that pneumonia risk can be progressively reduced by higher trough IgG levels, up to at least 1000 mg/DL [43]. Based on this meta-analysis, it was estimated that the risk of pneumonia is reduced by 27 % for each 100 mg/kg increase in trough IgG level. On average, trough IgG levels increase by 100–200 mg/DL for each 100 mg/kg increase in the IgG dose [38, 43, 44].

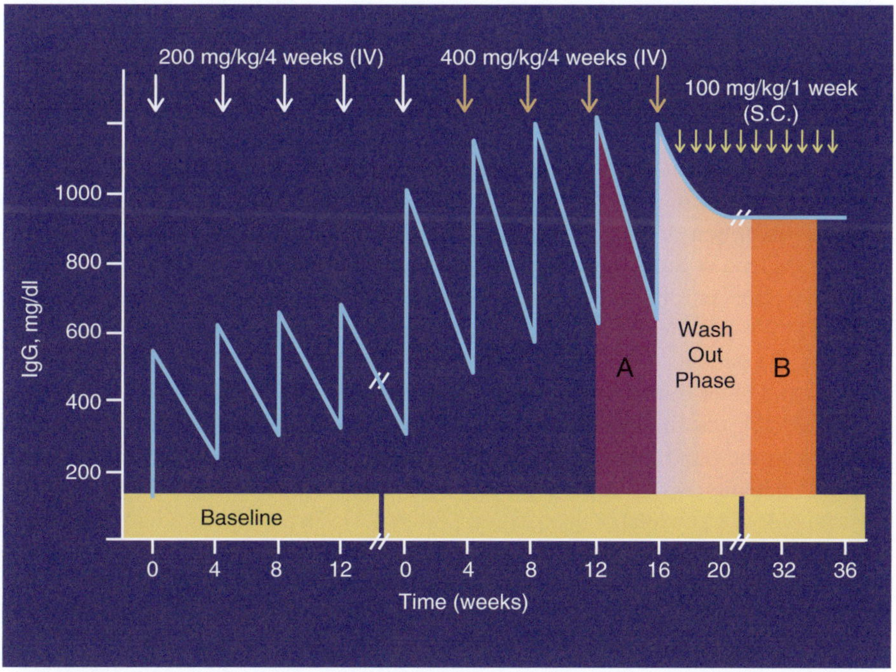

Fig. 4.2 IgG peak and trough levels depend on monthly dose, interval, and route of injection. At low monthly intravenous infusions (200 mg/kg), peak levels are low, trough levels barely above the base line, and patient protection is marginal. Higher monthly dosing results in increase in both peak and trough levels, eventually reaching equilibration. A switch from monthly intravenous infusions (400 mg/kg/months) to weekly subcutaneous injections (100 mg/kg/week) results in an IgG trough level that remains stable

While some patients remain infection free at lower monthly IgG doses and lower trough levels, others do not, demonstrating a need to optimize therapy for individual patients to avoid recurrent infections and progression to chronic lung disease. The consensus is that there is not a "magic" monthly dose or trough level that keeps a patient infection free. Variables include the "base IgG level" (determined before starting therapy), the interval between infusions, the route of infusion (IVIG versus SCIG), the presence of bronchiectasis, protein loss, or increased catabolic destruction, and other factors.

IVIG as replacement therapy requires venous access that is often painful and difficult to maintain. However, most treating physicians agree that IVIG should not be given by a central line (e.g., Portacath) that is maintained only for IVIG therapy. The usual starting dose is ~400 mg/kg/months, but may have to be increased up to 800–1000 mg/kg/months. Because adverse events are more frequent when the interval between infusions reaches 4 weeks, a 3-week schedule is preferred. Most patients who self-infuse IVIG at home prefer a 2-week interval to avoid "pre-infusion blues" and post-infusion discomfort. The rate of infusion tolerated differs from patient to patient and may depend on the preparation and the health status of the patient. "Rapid" infusion (over 400 mg/kg/h) is associated with increased AEs [45] and may lead to serious complications if preparations with high osmolality are used [34]. A select group of patients, who tolerated the infusions well and were hemodynamically stable, had been taught how to self-infuse intravenously or be infused by a family member at home, or receive IVIG at home by a trained infusion nurse, improving quality of life and patient health perception [46, 47].

SCIG replacement had a slow start [48–51], when, in the early 1980s, a few patients received Cohn fraction II by slow subcutaneous infusion. Presumably due to an outbreak of non-A non-B hepatitis [29] following treatment with an IVIG preparation produced in Stockholm (KabiVitrum) and its abrupt withdrawal, Gardulf and co-investigators enrolled 25 patients in a clinical trial using a pump to infuse rapidly up to 20 ml/h per site of a 16 % "intramuscular" preparation with excellent acceptance by the patients [52]. As a result, SCIG became the preferred Ig treatment in Sweden and other European countries. More recently, following clinical trials in the USA [53–55], the FDA approved several preparations for subcutaneous injection, resulting in the rapid increase in the use of SCIG in North America, especially for pediatric patients. As of 2015, a total of four preparations have been approved and are available for subcutaneous infusion (Table 4.1). Two of these, Gammagard and Gamunex, both 10 % IgG, are also licensed for intravenous use. Hizentra is a 20 % Ig preparation exclusively produced for subcutaneous injection. Most recently, the FDA licensed HyQvia, a "dual vial" preparation consisting of a vial of recombinant human hyaluronidase and a vial of 10 % IgG in glycine buffer, which is injected locally via the same needle as the subcutaneously injected hyaluronidase [56].

SCIG is ideally suited for self-infusion at home [57], can be given weekly, every other week, or in the case of HyQvia, every 3–4 weeks, using a pump or by push. SCIG can be given in one or multiple sites, using the abdomen, lateral thighs, or arms. Infusing 1 ml/min per site is tolerated well by most patients, including infants, who might benefit from a cream-based local anesthetic, such as EMLA cream. Self-infusion at home provides individualized therapy that may consist of injection by push once weekly if the volume is small (as for infants), or more often (in older children or adults), or injection by pump. Frequent SC infusions result in high "trough levels" that remain relatively stable (Fig. 4.2). At any rate, self-infusion at home by IV or SC provides patients with more control of therapy and improves quality of life [46, 47, 55, 58].

Conclusion

Cohn fractionation of plasma, developed as part of the Second World War efforts, has provided the starting material for safe and effective immunoglobulin preparations which are available, after decades of development, for intravenous or subcutaneous injection, providing antibody-deficient patients with multiple choices to stay free of infections. Some will opt for periodic visits at infusion centers, while others may prefer self-infusion at home—preferably every 2–3 weeks if using IVIG, or weekly, daily, or monthly if using SCIG. There are those who prefer a simple spring-loaded pump (Freedoms-60) and those who cannot resist using complex electronic pumps. Some follow the Swedish model and use a $0.55 butterfly needle that can be bent to ensure the tip is in the subcutaneous compartment; others prefer expensive tubing, with the needle mounted at a 90° angle to a plastic circle (at $12.00 a set). Some learn the hard way that if the needle is too short, part of the material leaks and goes into the intradermal space, hurting locally, or if the needle is too long, the material is injected intramuscularly, resulting in local soreness that can last days, similar to gammaglobulin preparations designed to be given intramuscularly. Experienced self-infusers have learned that a "wet" needle, resulting from priming the tubing, causes intradermal deposit of IgG—and pain, some have learned to adjust the dose as needed—more when sick or during the cold season, less when healthy or during summer. And, unfortunately, some tend to postpone or forget to infuse—demonstrating that they are not well suited for self-infusion.

References

1. Ochs HD (2014) Immunity: from serendipitous observations to science-based specialty. In: Etzioni A, Ochs HD (eds) Primary immunodeficiency disorders, a historic and scientific perspective. Academic, Oxford, pp 1–11
2. Behring E, Kitasato S (1890) Über das Zustandekommen der Diphtherie-Immunität und der Tetanus-Immunität bei Thieren. Dtsch Med Wschr 16:1113–1114, PMC
3. Behring E, Ehrlich P (1894) Zur Diphtherie Immunisierungs- und Heilungsfrage. Dtsch Med Wochenschr 20:437–438, PMC
4. Buchner H (1889) Über die nähere Natur der bacterientödtenden Substanz im Blutserum. Zentralbl Bakteriol 16:561–565, PMC
5. Heymann B (1928) Zur Geschichte der Seitenkettentheorie Paul Ehrlichs. Klin Wschr 7:1257–1260, 1305-1309. PMC
6. Ehrlich P (1900) On immunity with special reference to cell life. Croonian lecture. Proc R Soc 66:424–448, PMC
7. Wolff-Eisner A (1910) Handbuch der Serumtherapie und experimentellen Therapie: ein Handbuch für Klinik und Paxis. JF Lehmanns Verlag, München, PMC
8. Weisbecker L (1896) XV. Heilserum gegen Masern. Zeit f klin Med 30:312, PMC

9. Cenci F (1907) Alcune esperienze di sieroimmunizzazione e sieroterapia nel morbillo. Riv Clin Pediatr 5:1017–1025, PMC
10. McKhann CF, Chu FT (1933) Use of placental extract in prevention and modification of measles. Am J Dis Child 45:475–479, PMC
11. Karelitz S (1938) Prophylaxis against measles with the globulin fraction of immune human adult serum. Am J Dis Child 55:768–775, PMC
12. Tiselius A, Kabat EA (1939) An electrophoretic study of immune sera and purified antibody preparations. J Exp Med 69(1):119–131, PMC 2133726
13. Cohn EJ, Luetscher JA, Oncley JL, Armstrong SH, Davis BD (1940) Preparation and properties of serum and plasma proteins. III. Size and charge of proteins separating upon equilibration across membranes with ethanol-water mixtures of controlled pH, ionic strength and temperature. J Am Chem Soc 62:3396–3400, PMC
14. Cohn EJ, Oncley JL, Strong LE, Hughes WL, Armstrong SH (1944) Chemical, clinical, and immunological studies on the products of human plasma fractionation. I. The characterization of the protein fractions of human plasma. J Clin Invest 23(4):417–432, PMC 435357
15. Enders JF (1944) Chemical, clinical, and immunological studies on the products of human plasma fractionation. X. The concentrations of certain antibodies in globulin fractions derived from human blood plasma. J Clin Invest 23(4):510–530, PMC 435366
16. Ordman CW, Jennings CG, Janeway CA (1944) Chemical, clinical, and immunological studies on the products of human plasma fractionation. XII. The use of concentrated normal human serum gamma globulin (human immune serum globulin) in the prevention and attenuation of measles. J Clin Invest 23(4):541–549, PMC 435368
17. Stokes J, Maris EP, Gellis SS (1944) Chemical, clinical, and immunological studies on the products of human plasma fractionation. XI. The use of concentrated normal human serum gamma globulin (human immune serum globulin) in the prophylaxis and treatment of measles. J Clin Invest 23(4):531–540, PMC 435367
18. Keller MA, Stiehm ER (2000) Passive immunity in prevention and treatment of infectious diseases. Clin Microbiol Rev 13(4):602–614, PMC 88952
19. Bruton OC (1952) Agammaglobulinemia. Pediatrics 9(6):722–728, PMC
20. Janeway CA, Apt L, Gitlin D (1953) Agammaglobulinemia. Trans Assoc Am Physicians 66:200–202, PMC
21. Sanford JP, Favour CB, Tribeman MS (1954) Absence of serum gamma globulins in an adult. N Engl J Med 250(24):1027–1029, PMC
22. Janeway CA (1970) The development of clinical uses of immunoglobulins: a review. In: Merler E (ed) Immunoglobulins. National Academy of Sciences, Washington, DC, pp 3–14
23. Barandun S, Kistler P, Jeunet F, Isliker H (1962) Intravenous administration of human gammaglobulin. Vox Sang 7:157–174, PMC
24. Barandun S, Morell A, Skvaril F (1979) Clinical experiences with immunoglobulin for intravenous use. In: Alving BM, Finlayson JS (eds) Immunoglobulins: characteristics and uses of intravenous preparations. Department of Health and Human Services, Bethesda, pp 31–35
25. Ammann AJ, Ashman RF, Buckley RH, Hardie WR, Krantmann HJ, Nelson J, Ochs H, Stiehm ER, Tiller T, Wara DW, Wedgwood R (1982) Use of intravenous gamma-globulin in antibody immunodeficiency: results of a multicenter controlled trial. Clin Immunol Immunopathol 22(1):60–67, PMC
26. Ochs HDBR, Pirofsky B, Fischer SF, Rousell RH, Anderson CJ, Wedgwood RJ (1980) Safety and patient acceptability of intravenous immune globulin in 10% maltose. Lancet ii 1980:1158–1159, PMC
27. Finlayson JS (1979) Overview of potential uses for imunoglobulin preparations, possible etiologies of adverse reactions, and ideal characteristics of intravenous preparations. In: Alving BM, Finlayson JS (eds) Immunoglobulins: characteristics and uses of intravenous preparations. Department of Health and Human Services, Bethesda, pp 229–233
28. Yap PL, McOmish F, Webster AD, Hammarstrom L, Smith CI, Bjorkander J, Ochs HD, Fischer SH, Quinti I, Simmonds P (1994) Hepatitis C virus transmission by intravenous immunoglobulin. J Hepatol 21(3):455–460, PMC

29. Bjoro K, Froland SS, Yun Z, Samdal HH, Haaland T (1994) Hepatitis C infection in patients with primary hypogammaglobulinemia after treatment with contaminated immune globulin. N Engl J Med 331(24):1607–1611, PMC
30. Bresee JS, Mast EE, Coleman PJ, Baron MJ, Schonberger LB, Alter MJ, Jonas MM, Yu MY, Renzi PM, Schneider LC (1996) Hepatitis C virus infection associated with administration of intravenous immune globulin. A cohort study. JAMA 276(19):1563–1567, PMC
31. Stiehm ER (2013) Adverse effects of human immunoglobulin therapy. Transfus Med Rev 27(3):171–178, PMC
32. Bonilla FA (2008) Intravenous immunoglobulin: adverse reactions and management. J Allergy Clin Immunol 122(6):1238–1239, PMC
33. Scribner CL, Kapit RM, Phillips ET, Rickles NM (1994) Aseptic meningitis and intravenous immunoglobulin therapy. Ann Intern Med 121(4):305–306, PMC
34. Centers for Disease Control and Prevention (CDC) (1999) Renal insufficiency and failure associated with immune globulin intravenous therapy--United States 1985-1998. MMWR Morb Mortal Wkly Rep 48(24):518–521, PMC
35. Roemisch JR, Kaar W, Zoechling A, Kannicht C, Putz M, Kohla G (2011) Identification of activated FXI as the major biochemical root cause in IVIG batches associated with thromboembolic events. Analytical and experimental approaches resulting in corrective and preventive measures implemented into the octagam® manufacturing process. WebmedCentral Immunotherapy 2(6):1–16, PMC
36. Moscow JA, Casper AJ, Kodis C, Fricke WA (1987) Positive direct antiglobulin test results after intravenous immune globulin administration. Transfusion 27(3):248–249, PMC
37. Daw Z, Padmore R, Neurath D, Cober N, Tokessy M, Desjardins D, Olberg B, Tinmouth A, Giulivi A (2008) Hemolytic transfusion reactions after administration of intravenous immune (gamma) globulin: a case series analysis. Transfusion 48(8):1598–1601, PMC
38. Lucas M, Lee M, Lortan J, Lopez-Granados E, Misbah S, Chapel H (2010) Infection outcomes in patients with common variable immunodeficiency disorders: relationship to immunoglobulin therapy over 22 years. J Allergy Clin Immunol 125(6):1354–1360.e1354, PMC
39. Shapiro RS, Wasserman RL, Bonagura V, Gupta S (2014) Emerging paradigm of primary immunodeficiency disease: individualizing immunoglobulin dose and delivery to enhance outcomes. J Clin Immunol 34(1):1–7, PMC
40. Roifman CM, Levison H, Gelfand EW (1987) High-dose versus low-dose intravenous immunoglobulin in hypogammaglobulinaemia and chronic lung disease. Lancet 1(8541):1075–1077, PMC
41. Liese JG, Wintergerst U, Tympner KD, Belohradsky BH (1992) High- vs low-dose immunoglobulin therapy in the long-term treatment of X-linked agammaglobulinemia. Am J Dis Child 146(3):335–339, PMC
42. Eijkhout HW, van Der Meer JW, Kallenberg CG, Weening RS, van Dissel JT, Sanders LA, Strengers PF, Nienhuis H, Schellekens PT (2001) The effect of two different dosages of intravenous immunoglobulin on the incidence of recurrent infections in patients with primary hypogammaglobulinemia. A randomized, double-blind, multicenter crossover trial. Ann Intern Med 135(3):165–174, PMC
43. Orange JS, Grossman WJ, Navickis RJ, Wilkes MM (2010) Impact of trough IgG on pneumonia incidence in primary immunodeficiency: a meta-analysis of clinical studies. Clin Immunol 137(1):21–30, PMC
44. Ochs HD, Fischer SH, Wedgwood RJ, Wara DW, Cowan MJ, Ammann AJ, Saxon A, Budinger MD, Allred RU, Rousell RH (1984) Comparison of high-dose and low-dose intravenous immunoglobulin therapy in patients with primary immunodeficiency diseases. Am J Med 76(3A):78–82, PMC
45. Gelfand EW, Hanna K (2006) Safety and tolerability of increased rate of infusion of intravenous immunoglobulin G, 10% in antibody-deficient patients. J Clin Immunol 26(3):284–290, PMC
46. Daly PB, Evans JH, Kobayashi RH, Kobayashi AL, Ochs HD, Fischer SH, Pirofsky B, Sprouse C (1991) Home-based immunoglobulin infusion therapy: quality of life and patient health perceptions. Ann Allergy 67(5):504–510, PMC

47. Kobayashi RH, Kobayashi AD, Lee N, Fischer S, Ochs HD (1990) Home self-administration of intravenous immunoglobulin therapy in children. Pediatrics 85(5):705–709, PMC
48. Roord JJ, van der Meer JW, Kuis W, de Windt GE, Zegers BJ, van Furth R, Stoop JW (1982) Home treatment in patients with antibody deficiency by slow subcutaneous infusion of gammaglobulin. Lancet 1(8273):689–690, PMC
49. Ugazio AG, Duse M, Re R, Mangili G, Burgio GR (1982) Subcutaneous infusion of gammaglobulins in management of agammaglobulinaemia. Lancet 1(8265):226, PMC
50. Welch MJ, Stiehm ER (1983) Slow subcutaneous immunoglobulin therapy in a patient with reactions to intramuscular immunoglobulin. J Clin Immunol 3(3):285–286, PMC
51. Berger M, Cupps TR, Fauci AS (1980) Immunoglobulin replacement therapy by slow subcutaneous infusion. Ann Intern Med 93(1):55–56, PMC
52. Gardulf A, Hammarstrom L, Smith CI (1991) Home treatment of hypogammaglobulinaemia with subcutaneous gammaglobulin by rapid infusion. Lancet 338(8760):162–166, PMC
53. Ochs HD, Gupta S, Kiessling P, Nicolay U, Berger M (2006) Safety and efficacy of self-administered subcutaneous immunoglobulin in patients with primary immunodeficiency diseases. J Clin Immunol 26(3):265–273, PMC
54. Wasserman RL, Irani AM, Tracy J, Tsoukas C, Stark D, Levy R, Chen J, Sorrells S, Roberts R, Gupta S (2010) Pharmacokinetics and safety of subcutaneous immune globulin (human), 10% caprylate/chromatography purified in patients with primary immunodeficiency disease. Clin Exp Immunol 161(3):518–526, PMC 2962970
55. Nicolay U, Kiessling P, Berger M, Gupta S, Yel L, Roifman CM, Gardulf A, Eichmann F, Haag S, Massion C, Ochs HD (2006) Health-related quality of life and treatment satisfaction in North American patients with primary immunedeficiency diseases receiving subcutaneous IgG self-infusions at home. J Clin Immunol 26(1):65–72, PMC
56. Wasserman RL (2014) Overview of recombinant human hyaluronidase-facilitated subcutaneous infusion of IgG in primary immunodeficiencies. Immunotherapy 6(5):553–567, PMC
57. Kobrynski L (2012) Subcutaneous immunoglobulin therapy: a new option for patients with primary immunodeficiency diseases. Biologics 6:277–287, PMC 3430092
58. Gardulf A, Nicolay U, Math D, Asensio O, Bernatowska E, Bock A, Costa-Carvalho BT, Granert C, Haag S, Hernandez D, Kiessling P, Kus J, Matamoros N, Niehues T, Schmidt S, Schulze I, Borte M (2004) Children and adults with primary antibody deficiencies gain quality of life by subcutaneous IgG self-infusions at home. J Allergy Clin Immunol 114(4):936–942, PMC

Genetic Variation in Bruton Tyrosine Kinase

5

Gerard C.P. Schaafsma and Mauno Vihinen

5.1 Introduction

Bruton agammaglobulinemia tyrosine kinase (*BTK*) variations lead to X-linked agammaglobulinemia (XLA, MIM# 300300), a hereditary primary immunodeficiency [26, 29]. XLA is caused by a block in B cell differentiation resulting in severely decreased numbers of B lymphocytes and an almost complete lack of plasma cells and very low or missing immunoglobulin levels of all isotypes. The patients have increased susceptibility to mainly bacterial infections because of virtually absent humoral immune responses. The frequency of XLA has been estimated to be 1:200,000 live births. The disease is considered to have full penetrance. Female carriers are healthy but display nonrandom X-chromosome inactivation in their B cells. Only a few female patients have been identified.

The *BTK* gene (LRG_128, reference sequence used U78027.1) contains 19 exons (Fig. 5.1) and codes for a protein of 77 kDa. Exon 1 is outside the coding region. BTK is expressed in all hematopoietic lineages except for T lymphocytes and plasma cells [23]. BTK belongs to the Tec family of related cytoplasmic protein tyrosine kinases (PTKs) formed by BMX (BMX non-receptor tyrosine kinase), ITK (IL2-inducible T cell kinase), TEC (tec protein tyrosine kinase), and TXK (TXK tyrosine kinase). Except for TXK, they have the same domain organization, from the N-terminus pleckstrin homology (PH) domain, Tec homology (TH) domain, Src homology 3 (SH3) domain, SH2 domain, and catalytic tyrosine kinase (TK) domain.

The three-dimensional structure has been determined for the PH domain and the first half of the TH domain [7], the SH3 domain [4], the SH2 domain [6], and the kinase domain [12]. For the full-length BTK, there is a low-resolution structure in an extended conformation [13]. BTK interacts with several partners [15].

G.C.P. Schaafsma • M. Vihinen (✉)
Department of Experimental Medical Science, Lund University,
Lund SE-22184, Sweden
e-mail: mauno.vihinen@med.lu.se

A. Plebani, V. Lougaris (eds.), *Agammaglobulinemia*,
Rare Diseases of the Immune System 4, DOI 10.1007/978-3-319-22714-6_5

Variations in BTK account for about 80 % of agammaglobulinemia cases. Several other genes can lead to a failure of B cell development and agammaglobulinemia [1]. These genes encode components of the pre-B cell receptor or proteins that are activated by cross-linking of the pre-B cell receptor. Defects in these genes lead to a block in B cell differentiation at the pro-B to pre-B cell transition. Other forms of agammaglobulinemia appear with growth hormone deficiency or as autosomal recessive diseases. Some autosomal recessive agammaglobulinemias have been identified involving pre-B cell receptor (pre-BCR) or BCR component genes μ-heavy chain (*IGHM*), λ5/14.1 (*IGLL1*, immunoglobulin lambda-like polypeptide 1), Igα (*CD79A*), and Igβ (*CD79B*). Variations in the B cell linker protein (*BLNK*), which is essential for Igμ signal transduction, and *PIK3R1* (phosphoinositide-3-kinase, regulatory subunit 1 (alpha)) for phosphoinositide 3-kinase regulator are downstream of BCR.

5.1.1 BTKbase

BTKbase is the first immunodeficiency variation database (IDbase) founded in 1994 [32]. Subsequently more than 130 immunodeficiency variation databases (IDbases) have been released [19]. BTKbase contains public variation entries for 1362 patients from 1198 unrelated families (total number of variants in these unrelated families is 1209) showing 742 unique molecular events.

BTKbase aims at collecting all published variations. Data are either directly submitted or derived from more than 100 publications. The database format has been previously published [31, 34]. The data are presented as individual entries, each carrying a unique patient identification number (PIN) and accession number, systematic names according to the Human Genome Variation Society (HGVS) variation nomenclature, a short verbal description of the variation, submission information (submission and update dates, version numbers, and submitter details), literature citations, and annotation in detail at DNA, RNA, and protein levels. In addition, the most important clinical parameters and laboratory findings are included, provided they are available.

IDbases, including BTKbase, follow a number of standards including the use of HUGO Gene Nomenclature Committee (HGNC) gene names (www.genenames.org), HGVS variation nomenclature [3], and IDRefSeqs (reference sequences for primary immunodeficiency genes and proteins). Currently, IDbases are in the process of changing to Locus Reference Genomic (LRG) reference sequences, which are already available for some 100 immunodeficiency genes (www.lrg-sequence.org). BTKbase follows the recommendations for locus-specific variation databases (LSDBs) [33] and their curation [2].

BTKbase is freely available at http://structure.bmc.lu.se/idbase/BTKbase/. The website contains information related to XLA and *BTK*. The bioinformatics pages include several tables for statistics of *BTK* variations. The variation distributions are shown along sequences in illustrative ways. The submission page provides variation checking facilities and electronic submission services. The variation browser allows visual means for browsing variations along the protein sequence. The reference information for variation publications and related protein structures are included in their own sections.

5.2 Analysis of BTK Variations

XLA arises as a block in B cell development. BTKbase contains information in many entries for the immunological status of patients. These properties have been extensively discussed in a previous publication [28]. The majority of the reported patients have significantly reduced numbers of B cells and Ig levels. A large portion of patients with X-linked diseases have de novo variations.

5.2.1 Variation Statistics

Extensive statistical analyses of variations at the three molecular levels, DNA, RNA, and protein, were performed. Since data per unique families are considered the most representative regarding, e.g., mutational effects and prevalence, the discussion about variation statistics mainly relates to these.

Variations appear throughout the BTK domains as well as in exons and introns (Fig. 5.1, Table 5.1); however, the distribution is not even. Some exons contain more variations than expected. The PH and SH2 domains contain approximately the expected number of variations, whereas there are less than expected in the TH and SH3 domains and more than expected in the kinase domain (Table 5.2). The TH domain has two structural elements [31, 35], an N-terminal BTK motif and a C-terminal proline-rich region which contains two proline-rich regions capable of intra- and intermolecular interactions [4, 17]. The reason for under-representativeness of the TH domain may be that it likely has a partially intrinsically disordered structure in the C-terminal half of the domain, and therefore, variations do not have a major effect. On the other hand, XLA-causing variants do appear in the Zn^{2+}-binding BTK motif.

We have recently investigated the putative effects of all possible amino acid substitutions due to single nucleotide changes in the BTK TK domain [27]. Altogether 67 % of the 1495 substitutions were predicted to be harmful. Although this number

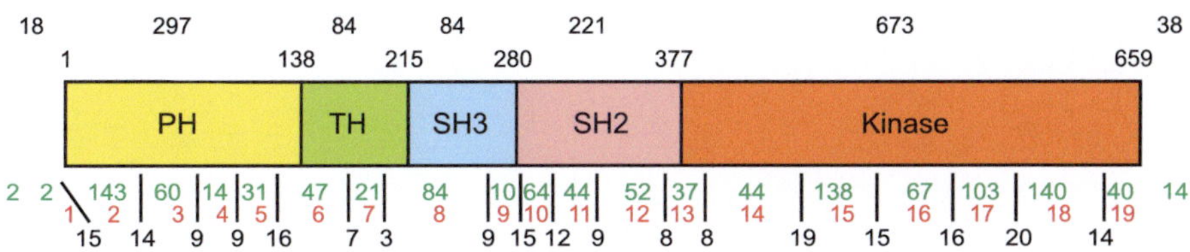

Fig. 5.1 Distribution of all variations to *BTK* gene regions and BTK protein domains. Variations in exons are indicated by green numbers in exons which are numbered in red. Variations in introns are in black below the domain chart. Domain borders are above the chart and numbers of variations in the domains above them

Table 5.1 Distribution of variation and variation types in BTK domains for all cases, independent families, and unique variations

Domain	Upstream			PH (414)			TH (231)			SH3 (194)			SH2 (292)			TK (846)			Others			Total			% of total		
Type	All	Fam	Uniq	All	Fam	Uniq	All	Fam	Uniq	All	Fam	Uniq	All	Fam	Uniq	All	Fam	Uniq	All	Fam	Uniq	All	Fam	Uniq	All	Fam	Uniq
Missense	0	0	0	119	100	48	11	10	10	0	0	0	112	93	42	328	290	147	0	0	0	570	493	247	41.5	40.8	33.3
Nonsense	0	0	0	44	40	19	16	14	9	46	39	11	23	22	12	97	85	43	0	0	0	226	200	94	16.4	16.5	12.7
Deletion; in-frame	1	1	1	11	6	6	2	2	2	0	0	0	5	5	4	21	18	11	0	0	0	40	32	24	2.9	2.6	3.2
Deletion; frameshift	0	0	0	3	2	1	0	0	0	0	0	0	1	1	1	2	1	1	4	3	3	10	7	6	0.7	0.6	0.8
Gross deletion	0	0	0	0	0	0	0	0	0	0	0	0	0	0	0	0	0	0	14	14	14	14	14	14	1.0	1.2	1.9
Insertion; in-frame	0	0	0	3	3	2	0	0	0	0	0	0	0	0	0	3	3	3	0	0	0	6	6	5	0.4	0.5	0.7
Insertion; frameshift	0	0	0	3	2	1	0	0	0	1	1	0	0	0	0	0	0	0	0	0	0	4	3	1	0.3	0.2	0.1
Indel	0	0	0	3	3	3	3	3	3	2	2	2	1	1	1	6	6	5	0	0	0	15	15	14	1.1	1.2	1.9
Intron	11	10	6	27	23	14	9	8	5	6	5	3	29	28	20	65	59	35	6	6	6	153	139	89	11.1	11.5	12.0
Intron; in-frame	0	0	0	3	3	2	0	0	0	1	1	1	6	4	3	0	0	0	1	1	1	11	9	7	0.8	0.7	0.9
Intron; out-of-frame	0	0	0	7	6	5	1	1	1	1	1	1	6	3	3	15	12	10	2	2	2	32	25	22	2.3	2.1	3.0
Frameshift	0	0	0	63	56	41	37	31	26	23	22	18	37	34	30	86	81	66	2	1	1	248	225	182	18.0	18.6	24.5
Multiple variant	0	0	0	7	6	4	5	5	5	2	2	2	0	0	0	6	4	2	0	0	0	20	17	13	1.5	1.4	1.8
Complex	2	1	1	3	2	2	0	0	0	2	2	2	1	1	1	2	2	2	1	1	1	11	9	9	0.8	0.7	1.2
Unknown	4	4	4	1	1	1	0	0	0	0	0	0	0	0	0	2	2	2	8	8	8	15	15	15	1.1	1.2	2.0
Total	18	16	12	297	253	149	84	74	61	84	75	40	221	192	117	633	563	327	38	36	36	1375	1209	742	100	100	100

Between brackets length of the domain in terms of cDNA nucleotides, *all* all analyzed alleles, *fam* unrelated families, *uniq* unique molecular events in DNA sequence

Table 5.2 Spectrum of variants in the structural BTK domains

Domain	Length	Length/total length	Normalized expected	Observed	χ^2	P^a
PH	414	0.209	242	253	0.47384	0.49122
TH	231	0.117	135	74	27.69467	$<10^{-6}$***
SH3	194	0.098	114	75	13.07900	0.00030***
SH2	292	0.148	171	192	2.60845	0.10630
TK	846	0.428	495	563	9.31070	0.00228**
Total	1977	1	1157	2257		

[a]Significance levels: ** $p<0.01$ *** $p<0.001$

seems very high, it is considered to be realistic because the kinase domain contains so many conserved regions and has several functions. The situation is likely very different in the SH3 and TH domains.

The variants are classified in Table 5.1 based on their effects on DNA or RNA level. The largest group of the variants is amino acid substitution causing missense variations (41 % of independent families). The SH3 domain is the only one where amino acid substitutions do not occur. Although SH3 domains are abundant in the human proteome, no disease-causing amino acid substitutions have been reported in any of them.

Nonsense variations account for 17 % of all variations, frameshift variations 19 %, and intronic variations 14 %. The proportions of deletions (4 %) and insertions (0.7 %) are very low and different from those reported in previous publications [10, 28] where proportions of 20 % (deletions) and 7 % (insertions) were given. These differences are due to the way the variants were counted, e.g., a variant with a DNA name "deletion" and an RNA name "frameshift" has been considered here as a frameshift variation. In the future we will avoid this kind of issues by adopting variation naming according to the Variation Ontology [30].

The distribution of variation types is very similar compared to the other IDbases [19]. The ratio of missense/nonsense variations, 2.5, is slightly higher in BTKbase compared to IDbases (1.5). Multiple variants in *BTK* have been identified in 17 families, complex variations in 9 families, and miscellaneous cases in 15 families.

There are altogether 341 unique amino acid substitutions. The theoretical maximum is 4151: thus, until now we have 8.2 % of the total variation; however, just a fraction of them is harmful and thus identifiable from XLA patients. In the case of nonsense variations, a larger portion has been seen in patients. There are 94 (28 %) of all the possible ($n=297$) variants in the BTKbase. According to χ^2 statistics, there is highly significant overrepresentation ($p<0.0001$).

When we are looking at the changes at amino acid level, it is apparent that arginine, as previously indicated, harbors the largest number of variants (Table 5.3). However, the most common outcome at protein level is protein truncation due to incorporation of a stop codon to the coding region. Altogether 29.5 % of single nucleotide changes lead to protein truncation.

Table 5.3 Amino acid substitutions indicated in percentages

									Hydrophilic													
	Hydrophobic								Acidic		Basic			Polar				Special				
->	A	F	I	L	M	V	W	Y	D	E	H	K	R	N	Q	S	T	C	G	P	X	Total
A	0					1			0.6	0.4						0	0		0	0.8	0	2.8
F		0	0	0.4		0.1		0.3								1		0.1			0.1	2.1
I		0.1	0	0	0.1	0						0	0	1	0	0.1	0.6				0	2
L		1.3	0.3	0	0	0.3	0.1				0		0.6		0.1	0.6				3.8	0.6	7.6
M			1.1	0.1	0	0.4						0.4	0.1				2				0	4.2
V	0.4	0.7	0	0	0	0.1			0.4	0.1									0.3	0	0.1	2.2
W				0.1									0.8			0.3		0.3	0		4.1	5.6
Y		0						0	0.7		0.6			0.6		0.8		1.3			4.8	8.7
D	0					0.4		0.1	0	0.1	0.1			0.3					0.3		0	1.4
E	0					0			0.7	0		0.3			0				0.4		1.8	3.2
H				0				0.1	0.1		0		0.4	0	0.1					0.3	0	1.1
K			0.1		0					1.1		0	0.4	0.3	0		0				1.4	3.4
R			0	0.3	0		6.3				4.1	0.6	0.1		4.5	1	0.3	2.5	2	1.1	8.3	31
N			0					0.1	0		0	0.1		0		0	0				0.1	0.4
Q			0	0						0	0.3	0	0.1		0					0.1	6.2	6.7
S	0	0.7	0.1	0.1			0	0.4					0.1	0		0	0	0	0	0.7	1	3.2
T	0.1		0.4		0							0	0	0		0	0			1.1	0	1.7
C		0.7					0.3	1.7					0.4			0.3		0	0.4		0.7	4.5
G	0.1					0.1	0.1		1.1	2.1			1.5			0		0	0.1		0.3	5.6
P	0.3			0.6		0					0		0.3		0	0.7	0.7			0	0	2.5
Total	1	3.5	2.1	1.7	0.1	2.5	6.9	2.8	3.6	3.9	5	1.4	5	2.1	4.8	4.8	3.5	4.2	3.5	8	29.5	100

Arginine is by far the most frequently substituted amino acid (31 %). This has not only been observed in BTK before [28] but also in variant datasets extracted from dbSNP [21], and this overrepresentation of arginine is known to be due to the high mutability of the codons containing CpG dinucleotides. Arginine is coded by six codons, four of which have a CpG dinucleotide in the first and second codon position [18]. The overrepresentation of arginine as the most frequently substituted amino acid also leads to the enrichment of tryptophan as the residue other amino acids are substituted to; arginine was replaced by tryptophan in 6.3 % of all amino acid substitutions (Table 5.3).

Proline is the amino acid to which most amino acids have been substituted to (8 %) closely followed by tryptophan (6.9 %), histidine (5 %), arginine (5 %), glutamine (4.8 %), and serine (4.8 %).

The G>A and C>T substitutions form the largest classes of changes, ~24 % (Table 5.4). The types of base changes were investigated more closely. The changes from amino to keto base and vice versa are much more frequent than substitutions within these groups. There is clearly a higher frequency of transitions (purine to purine and pyrimidine to pyrimidine, 66 %) than transversions (34 %). The higher rate of transitions agrees with the higher rate (~70 %) of transitions found to be typical for human genes [25]. The strong to weak base substitutions are by far the biggest category, containing 60 % of the variations. This was also found in the VariSNP variant datasets [21].

Table 5.4 Nucleotide substitutions in unique families (%)

All

→	a	c	g	t	Total
a	0	3.5	8.3	2.8	14.6
c	5	0	2.1	23.5	30.7
g	23.7	5.5	0	7.8	36.9
t	3.6	10	4.2	0	17.8
Total	32.3	19	14.6	34.1	100

Summarized into amino and keto categories

→	Amino	Keto	Total
Amino	8.5	36.7	45.2
Keto	42.8	12	54.8
Total	51.3	48.7	100

Summarized into weak and strong categories

→	Weak	Strong	Total
Weak	6.4	26	32.4
Strong	60	7.6	67.6
Total	66.4	33.6	100

Summarized into purine and pyrimidine categories

→	Purines	Pyrimidines	Total
Purines	31.9	19.6	51.5
Pyrimidines	14.9	33.6	48.5
Total	46.9	53.1	100

5.2.2 Structural Consequences

BTK consists of five domains which, except for the SH3 domain, contain amino acid substitutions (Fig. 5.2). The effects and consequences of the variations vary widely. A recent study revealed that about two thirds of all kinase domain variations originating from a single nucleotide change likely lead to XLA [27]. This is not to say that two thirds of all possible amino acid changes were harmful since the majority of them do not originate from single base changes (because of the organization of the genetic code). Numerous variants affect functional sites, such as ligand- and substrate-binding regions at the domains. Stability-affecting changes are common. There are putative explanations available for the consequences of all the 1495 substitutions studied. These results are well in line with previous studies and predictions of BTK variants [5, 7–9, 11, 12, 14, 16, 20, 22, 24, 28, 32, 35–41].

Minor changes can be accommodated without major structural alterations. As has been seen in especially the PH domain, changes to electrostatics are common [16]. When the charge is reversed, added, or removed, the properties of the site are modified. If this happens on the protein surface of the binding site, then the interactions with partners are impaired or weakened.

Structural variations appear frequently in secondary structural elements. Although there are some variations at loops connecting these elements, the α- and

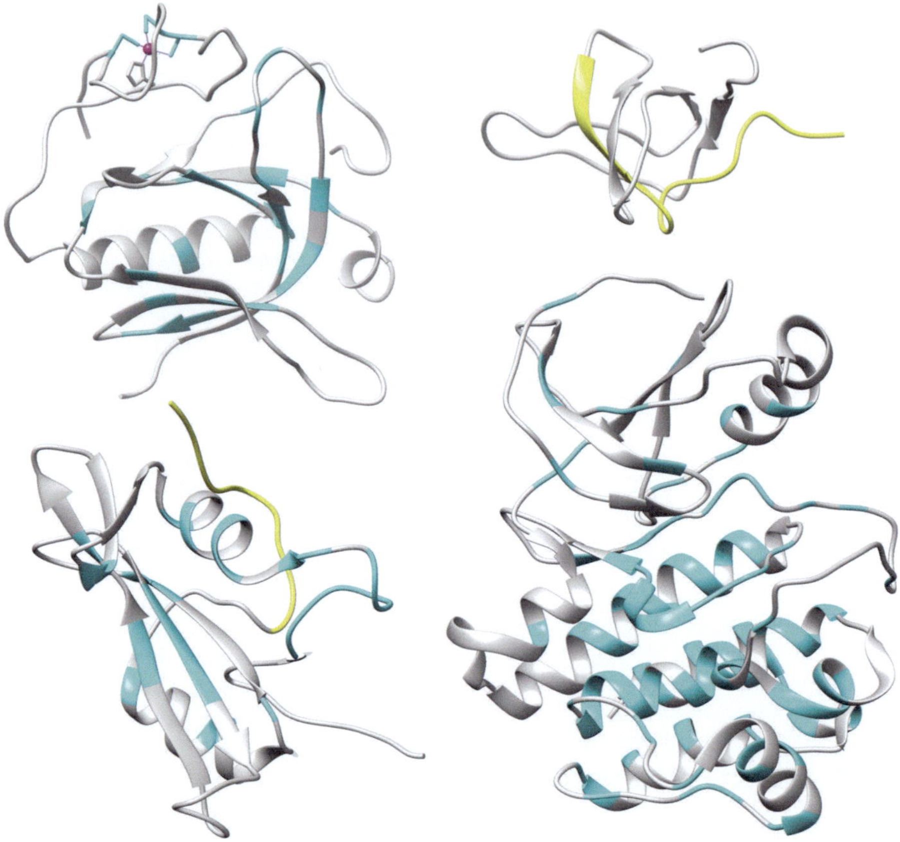

Fig. 5.2 Distribution of amino acid substitutions to BTK domains. Affected amino acids are shown in yellow. PH domain is on top left (PDB code 1BTK [7]). The first part of the TH domain including the BTK motif binding Zn^{2+} (magenta) is on the top of the domain. In the SH3 (1AWX [4]), top right, and SH2 domain (2GE9 [6]), bottom left, an in-frame deletion of 21 residues is indicated. The kinase domain (1K2P [12]) is at *bottom to the right*. Amino acid substitutions appear throughout all the domains except the SH3 domain where there are none

β-structures are more sensitive for substitutions. Structural variants are frequent on the protein core where there is no space for larger side chains due to tight packing. Further, introduction of charged or polar residues to the protein core, even if sterically possible, is usually harmful. Much more variation is allowed on the protein surface in areas not involved in intra- or intermolecular interactions. Some of these interactions are known; however, we do not even know the three-dimensional organization of the entire BTK. The domains are independently folding and connected by loops, which can be quite long. It is likely that the domain interactions are different in different folds of the entire protein. There is structural information for the entire BTK in elongated conformation [13]; however, this conformation is not likely the only one.

BTK variation information has been collected already for two decades into BTKbase, which has been a central resource for research and diagnosis. The database is constantly growing; however, the recent explosion in sequencing activities has not contributed much to the increased numbers in the database. That is presumably because many cases remain in laboratories and are never published or submitted to a database. It is in the interest of the entire community to share information about variations, especially in rare diseases such as XLA.

Acknowledgments Financial support from Lund University, Medical Faculty, and Swedish Research Council Vetenskapsrådet is gratefully acknowledged.

References

1. Berglöf A, Turunen JJ, Gissberg O, Bestas B, Blomberg KE, Smith CI (2013) Agammaglobulinemia: causative mutations and their implications for novel therapies. Expert Rev Clin Immunol 9:1205–1221
2. Celli J, Dalgleish R, Vihinen M, Taschner PE, den Dunnen JT (2012) Curating gene variant databases (LSDBs): toward a universal standard. Hum Mutat 33:291–297
3. den Dunnen JT, Antonarakis SE (2000) Mutation nomenclature extensions and suggestions to describe complex mutations: a discussion. Hum Mutat 15:7–12
4. Hansson H, Mattsson PT, Allard P, Haapaniemi P, Vihinen M, Smith CIE, Härd T (1998) Solution structure of the SH3 domain from Bruton tyrosine kinase. Biochemistry 37:2912–2924
5. Holinski-Feder E, Weiss M, Brandau O, Jedele KB, Nore B, Bäckesjö CM, Vihinen M, Hubbard SR, Belohradsky BH, Smith CIE, Meindl A (1998) Mutation screening of the BTK gene in 56 families with X-linked agammaglobulinemia (XLA): 47 unique mutations without correlation to clinical course. Pediatrics 101:276–284
6. Huang KC, Cheng HT, Pai MT, Tzeng SR, Cheng JW (2006) Solution structure and phosphopeptide binding of the SH2 domain from the human Bruton tyrosine kinase. J Biomol NMR 36:73–78
7. Hyvönen M, Saraste M (1997) Structure of the PH domain and Btk motif from Bruton tyrosine kinase: molecular explanations for X-linked agammaglobulinaemia. Embo J 16:3396–3404
8. Jin H, Webster ADB, Vihinen M, Sideras P, Vořechovský I, Hammarström L, Bernatowska-Matuszkiewicz E, Smith CIE, Bobrow M, Vetrie D (1995) Identification of Btk mutations in 20 unrelated patients with X-linked agammaglobulinaemia (XLA). Hum Mol Genet 4:693–700
9. Korpi M, Väliaho J, Vihinen M (2000) Structure-function effects in primary immunodeficiencies. Scand J Immunol 52:226–232
10. Lindvall JM, Blomberg KE, Väliaho J, Vargas L, Heinonen JE, Berglöf A, Mohamed AJ, Nore BF, Vihinen M, Smith CIE (2005) Bruton tyrosine kinase: cell biology, sequence conservation, mutation spectrum, siRNA modifications, and expression profiling. Immunol Rev 203:200–215
11. Maniar HS, Vihinen M, Webster ADB, Nilsson L, Smith CIE (1995) Structural basis for X-linked agammaglobulinemia (XLA): mutations at interacting Btk residues R562, W563, and A582. Clin Immunol Immunopathol 76:S198–S202
12. Mao C, Zhou M, Uckun FM (2001) Crystal structure of Bruton tyrosine kinase domain suggests a novel pathway for activation and provides insights into the molecular basis of X-linked agammaglobulinemia. J Biol Chem 276:41435–41443
13. Márquez JA, Smith CIE, Petoukhov MV, Lo Surdo P, Mattsson PT, Knekt M, Westlund A, Scheffzek K, Saraste M, Svergun DI (2003) Conformation of full-length Bruton tyrosine kinase (Btk) from synchrotron X-ray solution scattering. Embo J 22:4616–4624

14. Mattsson PT, Lappalainen I, Bäckesjö CM, Brockmann E, Lauren S, Vihinen M, Smith CIE (2000) Six X-linked agammaglobulinemia-causing missense mutations in the Src homology 2 domain of Bruton tyrosine kinase: phosphotyrosine-binding and circular dichroism analysis. J Immunol 164:4170–4177
15. Mohamed AJ, Yu L, Bäckesjö CM, Vargas L, Faryal R, Aints A, Christensson B, Berglöf A, Vihinen M, Nore BF and others (2009) Bruton tyrosine kinase (Btk): function, regulation, and transformation with special emphasis on the PH domain. Immunol Rev 228:58–73
16. Okoh MP, Vihinen M (1999) Pleckstrin homology domains of tec family protein kinases. Biochem Biophys Res Commun 265:151–157
17. Okoh MP, Vihinen M (2002) Interaction between Btk TH and SH3 domain. Biopolymers 63:325–334
18. Ollila J, Lappalainen I, Vihinen M (1996) Sequence specificity in CpG mutation hotspots. FEBS Lett 396:119–122
19. Piirilä H, Väliaho J, Vihinen M (2006) Immunodeficiency mutation databases (IDbases). Hum Mutat 27:1200–1208
20. Saha BK, Curtis SK, Vogler LB, Vihinen M (1997) Molecular and structural characterization of five novel mutations in the Bruton tyrosine kinase gene from patients with X-linked agammaglobulinemia. Mol Med 3:477–485
21. Schaafsma GCP, Vihinen M (2015) VariSNP, a benchmark database for variations from dbSNP. Hum Mutat 36:161–166
22. Shen B, Vihinen M (2004) Conservation and covariance in PH domain sequences: physicochemical profile and information theoretical analysis of XLA-causing mutations in the Btk PH domain. Protein Eng Des Sel 17:267–276
23. Smith CIE, Baskin B, Humire-Greiff P, Zhou JN, Olsson PG, Maniar HS, Kjellén P, Lambris JD, Christensson B, Hammarström L, Bentley D, Vetrie D et al (1994) Expression of Bruton agammaglobulinemia tyrosine kinase gene, BTK, is selectively down-regulated in T lymphocytes and plasma cells. J Immun 152:557–565
24. Speletas M, Kanariou M, Kanakoudi-Tsakalidou F, Papadopoulou-Alataki E, Arvanitidis K, Pardali E, Constantopoulos A, Kartalis G, Vihinen M, Sideras P, Ritis K (2001) Analysis of Btk mutations in patients with X-linked agammaglobulinaemia (XLA) and determination of carrier status in normal female relatives: a nationwide study of Btk deficiency in Greece. Scand J Immunol 54:321–327
25. Stephens JC, Schneider JA, Tanguay DA, Choi J, Acharya T, Stanley SE, Jiang R, Messer CJ, Chew A, Han JH and others (2001) Haplotype variation and linkage disequilibrium in 313 human genes. Science 293:489–493
26. Tsukada S, Saffran DC, Rawlings DJ, Parolini O, Allen RC, Klisak I, Sparkes RS, Kubagawa H, Mohandas T, Quan S, Belmont JW, Cooper MD et al (1993) Deficient expression of a B cell cytoplasmic tyrosine kinase in human X-linked agammaglobulinemia. Cell 72:279–290
27. Väliaho J, Faisal I, Ortutay C, Smith CIE, Vihinen M (2015) Characterization of all possible single nucleotide change –caused amino acid substitutions in the kinase domain of Bruton tyrosine kinase. Hum Mutat 36:638–647
28. Väliaho J, Smith CIE, Vihinen M (2006) BTKbase: the mutation database for X-linked agammaglobulinemia. Hum Mutat 27:1209–1217
29. Vetrie D, Vořechovský I, Sideras P, Holland J, Davies A, Flinter F, Hammarström L, Kinnon C, Levinsky R, Bobrow M, Smith CIE, Bentley DR (1993) The gene involved in X-linked agammaglobulinaemia is a member of the src family of protein-tyrosine kinases. Nature 361:226–233
30. Vihinen M (2014) Variation Ontology for annotation of variation effects and mechanisms. Genome Res 24:356–364
31. Vihinen M, Belohradsky BH, Haire RN, Holinski-Feder E, Kwan SP, Lappalainen I, Lehväslaiho H, Lester T, Meindl A, Ochs HD, Ollila J, Vořechovský I et al (1997) BTKbase, mutation database for X-linked agammaglobulinemia (XLA). Nucleic Acids Res 25:166–171
32. Vihinen M, Cooper MD, de Saint Basile G, Fischer A, Good RA, Hendriks RW, Kinnon C, Kwan SP, Litman GW, Notarangelo LD, Ochs HD, Rosen FS et al (1995) BTKbase: a database of XLA-causing mutations. Immunol Today 16:460–465

33. Vihinen M, den Dunnen JT, Dalgleish R, Cotton RG (2012) Guidelines for establishing locus specific databases. Hum Mutat 33:298–305
34. Vihinen M, Lehväslaiho H, Cotton RDH (1999) Immunodeficiency mutation databases. In: Ochs HD, Smith CIE, Puck JM (eds) Primary immunodeficiency diseases. A molecular and genetic approach. Oxford University Press, New York/Oxford, pp 443–447
35. Vihinen M, Nilsson L, Smith CIE (1994) Structural basis of SH2 domain mutations in X-linked agammaglobulinemia. Biochem Biophys Res Commun 205:1270–1277
36. Vihinen M, Nilsson L, Smith CI (1994) Tec homology (TH) adjacent to the PH domain. FEBS Lett 350:263–265
37. Vihinen M, Nore BF, Mattsson PT, Backesjo CM, Nars M, Koutaniemi S, Watanabe C, Lester T, Jones A, Ochs HD, Smith CI (1997) Missense mutations affecting a conserved cysteine pair in the TH domain of Btk. FEBS Lett 413:205–210
38. Vihinen M, Vetrie D, Maniar HS, Ochs HD, Zhu Q, Vořechovský I, Webster ADB, Notarangelo LD, Nilsson L, Sowadski JM, Smith CIE (1994) Structural basis for chromosome X-linked agammaglobulinemia: a tyrosine kinase disease. Proc Natl Acad Sci U S A 91:12803–12807
39. Vihinen M, Zvelebil MJ, Zhu Q, Brooimans RA, Ochs HD, Zegers BJ, Nilsson L, Waterfield MD, Smith CIE (1995) Structural basis for pleckstrin homology domain mutations in X-linked agammaglobulinemia. Biochemistry 34:1475–1481
40. Vořechovský I, Vihinen M, de Saint Basile G, Honsova S, Hammarström L, Müller S, Nilsson L, Fischer A, Smith CIE (1995) DNA-based mutation analysis of Bruton tyrosine kinase gene in patients with X-linked agammaglobulinaemia. Hum Mol Genet 4:51–58
41. Zhu Q, Zhang M, Rawlings DJ, Vihinen M, Hagemann T, Saffran DC, Kwan SP, Nilsson L, Smith CIE, Witte ON, Chen S-H, Ochs HD (1994) Deletion within the Src homology domain 3 of Bruton tyrosine kinase resulting in X-linked agammaglobulinemia (XLA). J Exp Med 180:461–470

Novel Therapeutic Options for X-Linked Agammaglobulinemia

6

Frank J.T. Staal

6.1 Introduction

A complex set of monogenetic diseases, known as primary immunodeficiencies (PIDs), are characterized by lack of functional immune cells or immune-cell-derived proteins [1–3]. PIDs are inherited disorders of the immune system, which as a group comprise more than 150 different monogenic conditions which predispose affected individuals to infections, allergy, autoimmune reactions, and cancer. Among the approximately 150 genes associated with PID that have been identified, two major categories are defects in B and/or T lymphocytes:

- Antibody deficiencies due to problems in B cell differentiation in the bone marrow or peripheral effector cell maturation (the so-called agamma-/hypogammaglobulinemias, in the case of peripheral B cell defects).
- Severe combined immunodeficiencies (SCID), constituting either isolated T cell or NK and B cell defects. In isolated T cell deficiencies, B cell function is hampered due to the lack of T cell help.

Among the B cell disorders originating in the bone marrow, the agammaglobulinemias are the most common, and among these, 85 % is covered by X-linked agammaglobulinemia (XLA). Other B cell disorders include the hyper IgM syndromes and some more rare deficiencies (Table 6.1). Taking all PIDs into account, X-linked agammaglobulinemia (XLA) is one of the most common primary immunodeficiencies in man and caused by mutations in Bruton's tyrosine kinase (BTK)

F.J.T. Staal
Department of Immunohematology and Blood Bank, L3-35, Leiden University Medical Center, Albinusdreef 2, P.O. Box 9600, Leiden 2300 RC, The Netherlands
e-mail: f.j.t.staal@lumc.nl

A. Plebani, V. Lougaris (eds.), *Agammaglobulinemia*,
Rare Diseases of the Immune System 4, DOI 10.1007/978-3-319-22714-6_6

Table 6.1 B cell defects and identified genes affected

Disease category	Known defects		Unknown
	Genes involved	Percentage	Percentage
Agammaglobulinemia	*BTK*	85	5–8 %
	IGHM	3–5	
	IGL14.1	<1	
	CD79A	<1	
	CD79B	<1	
	BLNK	<1	
Hyper IgM syndromes	*CD40L*	~50	25–35 %
	CD40	<1	
	AID	~15	
	UNG	<1	
	NEMO	~5	
Other BCR signaling defects	*CD19*	?	?
	CD21	?	?

Percentages are based on the European Society for Immunodeficiencies (ESID) database (www.ESID.org)

gene [4–12]. There are forms of common variable immunodeficiency (CVID) in which subclasses of IgGs and IgAs are deficient, which also are very common, but in many of these cases, the underlying gene defect is unknown. Although common variable immunodeficiency (CVID) is more common, this disease constitutes a spectrum of diseases underlying various genetic causes, many of which still are unknown. In contrast, the defective gene in XLA has been identified as a non-receptor protein tyrosine kinase, BTK, named after Dr. Bruton who first described the disease in boys lacking humoral immune responses. The BTK protein is homologous to the Src family of protein tyrosine kinases in that it contains Src homology domains. The major blockade caused by *BTK* mutations in human B cell differentiation is at the transition of pro-B cells into large pre-B1 cells. This results in accumulation of pro-B cells and severe reduction of pre-B1 cells in the bone marrow. In this transitional stage, pre-B-cell receptor signaling regulates clonal expansion. As BTK is associated with the pre-BCR signaling pathway, absence of *BTK* due to mutation leads to disruption of the pathway. In mice, absence of BTK will lead to X-linked immunodeficiency (Xid) phenotype. The Xid mice have a missense mutation at a conserved arginine residue within the pleckstrin (PH) domain of BTK, which results in conformational changes of the protein and subsequently lack of the ability to bind to the cell membrane [13]. In comparison to human XLA, the Xid phenotype is less severe due to redundancy of other Tec kinases, which apparently can compensate for the lack of BTK in murine B cells [14–17].

In this chapter, I will review the current therapeutic approaches used to treat XLA and the new possibilities offered for curative treatment through gene and cellular therapy.

6.2 Clinical Picture and Current Therapy

Patients with XLA usually present with a marked susceptibility to infections with encapsulated bacteria, particularly pneumococcus and *Haemophilus influenzae* type b. In addition to their increased susceptibility to these infections, patients with XLA have an elevated risk of infection with giardia, mycoplasma, and enteroviruses. Most infections are in the respiratory and gastrointestinal tract. Other bacterial infections, like cellulitis, arthritis, meningitis, and sepsis, have also been described, but are less common [5, 18, 19]. Giardia infections are associated with the development of inflammatory bowel disease. Infections with mycoplasma can result in chronic pneumonitis and/or arthritis. Viral infections do generally not cause unusual problems in patients with XLA as their T and NK cell number and function are normal; however, the enteroviruses, including echovirus, coxsackievirus, and wild-type and vaccine-associated poliovirus, can cause chronic, devastating enteroviral meningoencephalitis, arthritis, or enteritis [20].

XLA patients are unable to mount specific antibody responses and are lifelong and fully dependent on the specificity of the antibodies from the supplemented immunoglobulin preparations from healthy blood donors. Intravenous preparations became readily available in the late 1970s and mid-1980s. Nowadays, both intravenous and subcutaneous infusion allowed patients to receive higher doses of immunoglobulins that keep their serum IgG levels at, or near, that of normal individuals. Immunoglobulin supplementation in home care setting has become routine practice in many countries.

Affected boys generally begin to have frequent and protracted episodes of otitis and airway infections by 4–8 months of life after the passively transferred maternal antibodies declined. The average age XLA patients are recognized to have a defect in their immune system has fallen significantly over the last few decades, to the age of around 3 years nowadays [20, 21].

Earlier diagnosis allowed for earlier replacement therapy and increased attention to the diagnosis and therapy of their infections. Despite antibiotic prophylaxis and regular immunoglobulin substitution therapy, severe infections can emerge even in XLA patients with IgG trough levels within the normal ranges, possibly due to the fact that (secretory) IgA and IgM antibodies cannot be supplemented and remain absent. Higher IgG trough levels during immunoglobulin supplementation may lead to better protection [22, 23]. Nevertheless, even with optimal immunoglobulin supplementation, pediatric mortality remains high, with reported rates of 8.5–30 % depending on the length of observation [20].

Most boys with XLA remain well during the first 8 months of age by virtue of maternally transmitted IgG antibodies [6]. Female carriers of XLA are healthy and demonstrate no abnormalities of the immune system as B cells expressing the mutated gene are selected against [24].

Current treatment cannot be regarded as curative and will fail in case of serious infections that need highly specific antibodies. This results in severe life-threatening complications in young adulthood, but often also in children. In the older patients with XLA, progressive and irreversible organ damage develops. Particularly chronic respiratory tract infections may lead to debilitating lung disease, eventually making them candidates for lung transplantation. The hypogammaglobulinemia after lung transplantation confers an increased risk of infections and poorer survival. In addition, chronic ear infections leading to hearing impairment, increased risk of acquiring chronic infection with gastrointestinal parasites, and chronic enteroviral meningoencephalitis may all lead to poorer life quality. Chronic inflammatory reactions in mucosal surfaces predispose to an increased incidence of colorectal cancer [18].

Because of improvements in initial diagnosis, immunoglobulin replacement therapy, and treatment of infections, patients with XLA are less likely to develop long-term pulmonary insufficiency early in life and are expected to survive into early adulthood. Therefore, patients usually have to cope with this chronic disease during their further life span, which can severely reduce quality of life. Knowledge of the status of adults with XLA is important for caring for those patients, but is currently underdeveloped. However, less than 50 % of patients have been reported to reach the age of 45, often with very low quality of life due to severe lung complications [25] .

6.3 Introducing Gene Therapy

Gene therapy, using modified stem cells, is a possible curative lifelong treatment for XLA. The hematopoietic stem cell (HSC) is an adult-type, tissue-specific stem cell that gives rise to all blood cell lineages. Due to its long-lived nature and the key characteristic of self-renewal, the HSC is seen as an ideal cell for cell-based therapies aimed at reconstitution of the blood and immune system [26]. While most efforts have thus far been directed toward therapy with gene-corrected stem cells for SCID, XLA also is a good candidate.

Gene therapy uses autologous stem cells that have been treated ex vivo to replace or correct a deficient gene. For SCID, the current treatment of choice is allogeneic bone marrow transplantation (allo BMT). Although allo BMT carries significant risks, including mortality due to conditioning, graft failure, poor T cell immune reconstitution, and graft versus host disease (GvhD), this type of treatment has been tried in XLA patients before [27]. The fact that allo BMT has been tried underscores the severity of the disease. Unfortunately, the reported efforts were largely unsuccessful, most likely due to the lack of conditioning or minimal conditioning [27]. This most likely would not yield sufficient space in the bone marrow for proper donor reconstitution. It is conceivable that besides stem cell niches, also pre-B cell niches that are occupied by host BTK-deficient pre-B cells need to be cleared for this type of approach to be successful. Nevertheless, these attempts are important and point to the need for cellular replacement therapies in XLA, for instance, through gene therapy.

Several gene therapy trials have been conducted for different types of SCID, starting in the early 2000s [2, 3, 28–32]. In short, cells are harvested from the BM of the patient, in which a correct copy of the affected gene is inserted ex vivo and then given back to the patient. These studies used a mouse leukemia virus

(MLV)-based γ-retroviral vector to drive expression of the transgene by the long terminal repeat (LTR). A retrovirus reverse transcribes its RNA into DNA which then uses the enzyme integrase to integrate the DNA into the host genome. Due to this integration, the gene will be passed on to every daughter cell. Initially, there was restoration of functionality and patients demonstrated presence of cells from different lymphoid cell lineages in their peripheral blood together with improved immune functionality. Patients were able to go home and live in a normal environment and had normal growth and development. Unfortunately, thereafter it was reported that four patients in the X-SCID trial conducted in Paris and one patient in the London X-SCID trial out of the 20 total treated patients did develop T cell acute lymphoblastic leukemia (T-ALL) [31, 33, 34]. In the trial for ADA-SCID, which had used a similar vector [35], and in another ADA-SCID gene therapy trial conducted in London [36], no T-ALL development was observed. The T-ALLs in the X-SCID trials did result from insertional mutagenesis leading to ectopic expression of oncogenes. Below, the mechanisms of insertional mutagenesis and T-ALL development will be described in more detail. Due to the occurrence of these adverse side effects, new viral vectors were designed for delivery of the transgene (reviewed in [37, 38]).

The discovery that the therapeutic vector used for X-linked SCID gene therapy trial resulted in lymphoproliferations resembling T-ALL intensified discussions about the safety of gene therapy. More effort was put into improving safety of gene transfer procedures. Despite being a more complex virus, HIV-based vector systems might be more suitable for gene therapy. MLV and HIV have been shown to integrate in distinct regions within genes [39]. The tendency of MLV to integrate toward transcription start sites was seen as less favorable than the integration pattern of HIV, which does not display this partiality. Combined with built-in safety features, HIV-based vectors are thought to be a good alternative to MLV-based vectors for gene transfer.

6.4 Gene Therapy for XLA

Initial gene therapy efforts for XLA using the above-described mouse models made use of Moloney murine leukemia virus (MoLV)-based retroviral vectors (so-called γ−retroviral vectors). Of note, Yu and colleagues demonstrate that retroviral-mediated targeting of hematopoietic stem cells resulted in sustained correction of B cell development, rescue of serum immunoglobulin levels, and normalized B cell receptor-dependent proliferative functions in the BTK/Tec double-deficient mouse model, which reproduces the features of human XLA [40, 41]. Given the high risk of genotoxicity, the use of these vectors for PID gene therapy has been largely discontinued (reviewed in [38]). Instead, SIN LV vectors have been used. Three recent reports demonstrated that such vectors can be successfully used to develop gene therapy for XLA and demonstrated efficacy and safety in mouse models [42–44]. The principles of such gene therapy efforts are outlined in Fig. 6.1. In short, hematopoietic stem cells from the patient are transduced ex vivo with a lentiviral vector encoding the BTK gene. The gene is expressed either in all progeny of the stem cells

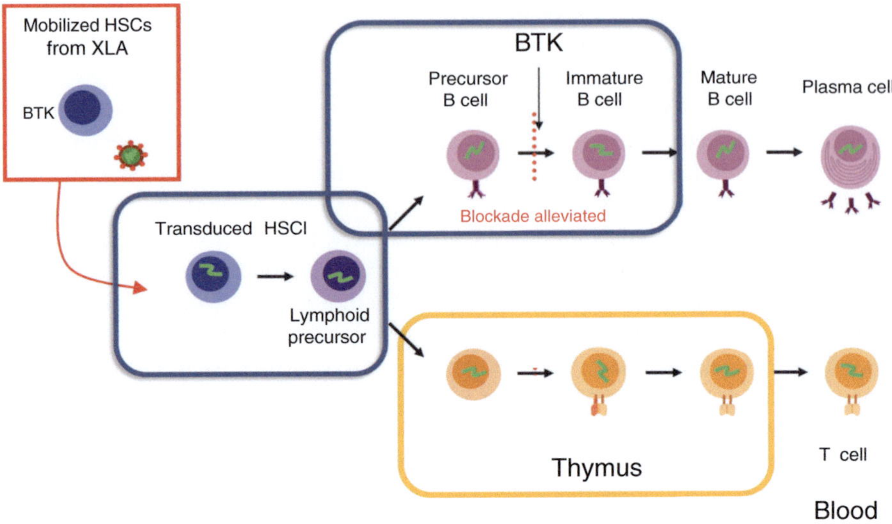

Fig. 6.1 Principle of stem cell-based gene therapy for XLA. HSCs from XLA patients are harvested from the bone marrow or mobilized peripheral blood and are transduced ex vivo with lentiviruses containing full-length human *BTK* gene. These gene-modified HSCs are transplanted back into the patients and the *BTK* gene is now correctly expressed, leading to alleviation of the block in pre-B cell development, maturation of B cells, and production of specific antibodies. Depending on the promoter used, the *BTK* gene can either be only expressed in the B cell lineage or also in other blood cells, for instance, in T cells, if broader promoters are used

or depending on the promoter used only in B lineage cells. The restored expression of the BTK protein rescues the pre-B cell development, resulting in normal B cells, which can differentiate into antibody-secreting plasma cells.

The successful production was reported in self-inactivating (SIN) lentiviral vectors expressing a codon-optimized version of the human *BTK* gene under control of three different ubiquitous or B cell-specific promoters [44]. These vectors were capable of restoring BTK protein levels in B lymphocytes to nearly normal levels. $BTK^{-/-}$ mice which were engrafted with transduced cells showed correction of both precursor B cell and peripheral B cell development. No correction was observed using the native *BTK* sequence, indicating the importance of codon optimization for clinical success. All treated mice exhibited the recovery of a special population of B cells, the B-1 cells in the peritoneal cavity. This B cell subpopulation with innate-like features is responsible for the formation of natural antibodies and is specifically absent in BTK-deficient mice [9, 15, 45, 46]. Importantly, serum IgM and IgG3 were restored to levels comparable to wild-type mice. Immunization of mice treated with corrected stem cells with the T cell-independent antigen TNP-Ficoll resulted in production of TNP-specific IgM and IgG3 antibodies, demonstrating the restoration of in vivo B cell responses. Calcium mobilization responses upon B cell receptor stimulation were restored indicating the B cell receptor signaling via tyrosine kinases was normal. Transplantation into secondary $BTK^{-/-}$ recipients resulted in full restoration of the peripheral B cell compartment and IgM levels. Moreover, no vector-related tumors were observed in secondary recipients when using the CD19 or EFS promoter [44].

Rawlings and coworkers used other B cell-specific promoters containing the GFP protein to identify and track transduced cells [42]. These investigators used a lentiviral vector (LV) containing the immunoglobulin enhancer (Eμ) and Igβ (B29) minimal promoter to drive B lineage-specific human BTK expression in $BTK^{-/-}Tec^{-/-}$ mice, a strain that reproduces the features of human XLA. Similar to our study, B cell-specific calcium fluxes were restored, and absolute numbers of mature B cell subsets in the PB, BM, and spleen and levels of Ig and B1 cells were found in nearly normal levels [42]. Again, secondary transplantations showed restored B cell development and lack of adversary effects as have been reported in the X-SCID trials [47].

In addition, Moreau and colleagues used a LV vector containing the CD19 promoter to drive BTK expression in human CD34+ cell B lymphoid progeny ex vivo [43].

An interesting point is to compare the studies by Ng et al. and Kerns et al. as both use LV vectors but with different promoters and different designs. In both studies BTK levels near wt levels (or at least more than half of wt) were required to obtain functional restoration. The Ng et al. study used codon optimization to obtain sufficient levels in the BTK-/- single-deficient mouse model, whereas in the Kerns et al. study, codon optimization was not required, most likely because the BTK-Tec double-deficient mice have a stronger block in precursor B cell development with an almost complete block at the pre-B cell stage, similar to human XLA. This will lead to a stronger positive selection of transduced B cells; therefore, codon optimization may not be required in these mice. On the other hand, initial attempts in the Ng et al. study also used the very strong SFFV promoter which showed substantial side effects due to development of erythromyeloid leukemias, most likely due to extremely high tyrosine kinase activity. As the expression levels with SFFV were several hundredfold higher than wild-type levels, the risk of development of these side effects is expected to be minor with the B cell-specific promoters used. Nevertheless, given the effects of vector variegation, there is still the risk that such extreme high BTK levels could occur in vivo in XLA patients treated with lentiviral vectors.

In conclusion, three different laboratories, two in Europe and one in the USA, have demonstrated the feasibility of using lentiviral vectors to restore BTK expression in murine models of XLA using gene-modified stem cells.

6.5 Gene Therapy by Splicing Correcting Oligonucleotides

In 15–20 % of the XLA patients, mutations in the BTK gene result from splice site mutations, in which the correct splicing of one or more exons is disrupted resulting in truncated or instable proteins [48]. In XLA, this can happen by mutations that introduce a new splice site that normally does not exist, leading to aberrant splicing. Several groups have taken the interesting approach of correcting these defects through splice site correction by administering short (15–25 nt) antisense oligonucleotides

(ASOs). ASOs are used to block the aberrant splice site in order to restore correct pre-mRNA splicing (reviewed in [49]). This has been demonstrated in mouse models in which BTK-deficient pre-B cells that carry conserved splice site mutations were treated with ASOs ex vivo and re-transplanted or in which the ASOs were delivered via transgenesis. A major challenge for this approach is how to target human pre-B cells in vivo or devise a strategy that would lead to long-term correction of BTK-deficient pre-B cells in vivo. The approach is also only amenable to the group of patients that have splice site mutations. Nevertheless, this constitutes a very elegant approach similar to approaches clinically tried in Duchenne muscular dystrophy [50].

6.6 Caveat: Risk Benefit Assessment for XLA Treatment via Gene Therapy

Current treatment for XLA includes subcutaneous or intravenous immunoglobulin (IVIG) infusion and prompt antibiotic administration to combat infections, which are only partially effective and not curative. IVIG and supportive treatment of XLA seem to be favorable for most patients during childhood. However, the patients are fully dependent on the specificity of the antibodies in the IVIG infusions and are unable to mount specific antibody responses. Therefore, current treatment cannot be regarded as curative. This explains why progressive and irreversible organ damage develops in these patients, resulting in severe life-threatening complications in young adulthood, but often also in children.

The complications lead to strongly reduced quality of life and are difficult to manage. Life expectancy for adults remains very poor. Given the high cost of IVIG treatment (>50,000 euros per year) and the high incidence of complications, gene therapy is clearly favorable. The risk-benefit ratio is always an important consideration for any new therapy. As SCID is a fatal disease whereas XLA is not, safety issues require more attention for this disease. The current state of the art with LV vectors however indicates a safety profile that allows for other PIDs to be considered as target diseases for gene therapy, as insertional mutagenesis risks are lower than with traditional non-SIN vectors, as recently demonstrated. Considering XLA, from a risk evaluation point of view, the pediatric cases with only initial and mild clinical signs of complications are among the first candidates for experimental therapy, in order to block the otherwise irreversible process of progressive organ damage. However, extreme overexpression of BTK which is difficult to reach with most viral vectors is associated with the risk of leukemogenesis. This is a disease-specific risk as BTK is an active tyrosine kinase which when expressed too high poses problems. As many precursor B cell leukemias are successfully treated with novel BTK inhibitors [51–56], the risk of BTK-induced leukemias by overexpression rather than xeno-toxicity should be considered. The risk is less with B cell-specific promoters, and indeed no such effects have been reported. To be on the safe side, novel developments in the gene-editing field, in particular using CRISP-CAS 9 system, are very encouraging. In such a system rather than adding an extra copy of the affected gene to the genome of patients (*gene addition*), the endogenous locus is repaired through homolog recombination. While such techniques for a long time were impossible due to the lack of proliferation of bona fide HSCs, important work by Naldini and coworkers [57] has demonstrated the feasibility of *gene editing* using such target-specific nucleases and novel ways to expand or proliferate HSCs ex vivo. For mutations in the IL2RG gene, the causative gene defect in X-SCID, HSCs could be corrected. In similar ways, the BTK gene or exons thereof could be replaced in XLA patients. This offers the advantage of controlled expression by the endogenous promoter. For BTK this would, for instance, also allow expression in the myeloid cell lineage, in which BTK also plays a functional role.

In conclusion, gene therapy for BTK deficiency has been proposed by various groups, but the currently used lentiviral vectors may pose too much of a risk for XLA gene therapy. However, encouraging new developments in gene editing offer high hopes for curative treatment for XLA through combination of stem cell-based gene therapy with gene-repaired HSCs.

References

1. Cavazzana-Calvo M, Hacein-Bey S, de Saint Basile G, Gross F, Yvon E, Nusbaum P, Selz F, Hue C, Certain S, Casanova JL et al (2000) Gene therapy of human severe combined immunodeficiency (SCID)-X1 disease. Science 288(5466):669–672
2. Fischer A (2004) Human primary immunodeficiency diseases: a perspective. Nat Immunol 5(1):23–30
3. Gaspar HB, Thrasher AJ (2005) Gene therapy for severe combined immunodeficiencies. Expert Opin Biol Ther 5(9):1175–1182
4. Aoki Y, Isselbacher KJ, Cherayil BJ, Pillai S (1994) Tyrosine phosphorylation of Blk and Fyn Src homology 2 domain-binding proteins occurs in response to antigen-receptor ligation in B cells and constitutively in pre-B cells. Proc Natl Acad Sci U S A 91(10):4204–4208
5. Conley ME, Fitch-Hilgenberg ME, Cleveland JL, Parolini O, Rohrer J (1994) Screening of genomic DNA to identify mutations in the gene for Bruton's tyrosine kinase. Hum Mol Genet 3(10):1751–1756
6. Conley ME, Rohrer J, Rapalus L, Boylin EC, Minegishi Y (2000) Defects in early B-cell development: comparing the consequences of abnormalities in pre-BCR signaling in the human and the mouse. Immunol Rev 178:75–90
7. de Weers M, Brouns GS, Hinshelwood S, Kinnon C, Schuurman RK, Hendriks RW, Borst J (1994) B-cell antigen receptor stimulation activates the human Bruton's tyrosine kinase, which is deficient in X-linked agammaglobulinemia. J Biol Chem 269(39):23857–23860
8. de Weers M, Verschuren MC, Kraakman ME, Mensink RG, Schuurman RK, van Dongen JJ, Hendriks RW (1993) The Bruton's tyrosine kinase gene is expressed throughout B cell differentiation, from early precursor B cell stages preceding immunoglobulin gene rearrangement up to mature B cell stages. Eur J Immunol 23(12):3109–3114
9. Hendriks RW, de Bruijn MF, Maas A, Dingjan GM, Karis A, Grosveld F (1996) Inactivation of Btk by insertion of lacZ reveals defects in B cell development only past the pre-B cell stage. Embo J 15(18):4862–4872
10. Middendorp S, Dingjan GM, Hendriks RW (2002) Impaired precursor B cell differentiation in Bruton's tyrosine kinase-deficient mice. J Immunol 168(6):2695–2703
11. Spaargaren M, Beuling E, Rurup ML, Meijer HP, Klok MD, Middendorp S, Hendriks RW, Pals S (2003) The B-cell antigen receptor controls integrin activity through Btk and PLCgamma2. J Exp Med 198(10):1539–1550

12. Middendorp S, Zijlstra AJ, Kersseboom R, Dingjan GM, Jumaa H, Hendriks RW (2005) Tumor suppressor function of Bruton tyrosine kinase is independent of its catalytic activity. Blood 105(1):259–265
13. Rawlings DJ, Saffran DC, Tsukada S, Largaespada DA, Grimaldi JC, Cohen L, Mohr RN, Bazan JF, Howard M, Copeland NG et al (1993) Mutation of unique region of Bruton's tyrosine kinase in immunodeficient XID mice. Science 261(5119):358–361
14. Kerner JD, Appleby MW, Mohr RN, Chien S, Rawlings DJ, Maliszewski CR, Witte ON, Perlmutter RM (1995) Impaired expansion of mouse B cell progenitors lacking Btk. Immunity 3(3):301–312
15. Khan WN, Alt FW, Gerstein RM, Malynn BA, Larsson I, Rathbun G, Davidson L, Muller S, Kantor AB, Herzenberg LA et al (1995) Defective B cell development and function in Btk-deficient mice. Immunity 3(3):283–299
16. Wicker LS, Scher I (1986) X-linked immune deficiency (xid) of CBA/N mice. Curr Top Microbiol Immunol 124:87–101
17. Witkowski J, Forrester LM, Ansell JD, Micklem HS (1985) Influence of the xid mutation on B lymphocyte development in adult mice. Adv Exp Med Biol 186:47–55
18. Conley ME, Howard V (2002) Clinical findings leading to the diagnosis of X-linked agammaglobulinemia. J Pediatr 141(4):566–571
19. Conley ME, Puck JM (1988) Carrier detection in typical and atypical X-linked agammaglobulinemia. J Pediatr 112(5):688–694
20. Winkelstein JA, Marino MC, Lederman HM, Jones SM, Sullivan K, Burks AW, Conley ME, Cunningham-Rundles C, Ochs HD (2006) X-linked agammaglobulinemia: report on a United States registry of 201 patients. Medicine (Baltimore) 85(4):193–202
21. Fiorini M, Franceschini R, Soresina A, Schumacher RF, Ugazio AG, Rossi P, Plebani A, Notarangelo LD (2004) BTK: 22 novel and 25 recurrent mutations in European patients with X-linked agammaglobulinemia. Hum Mutat 23(3):286
22. Bayrakci B, Ersoy F, Sanal O, Kilic S, Metin A, Tezcan I (2005) The efficacy of immunoglobulin replacement therapy in the long-term follow-up of the B-cell deficiencies (XLA, HIM, CVID). Turk J Pediatr 47(3):239–246
23. Kanegane H, Futatani T, Wang Y, Nomura K, Shinozaki K, Matsukura H, Kubota T, Tsukada S, Miyawaki T (2001) Clinical and mutational characteristics of X-linked agammaglobulinemia and its carrier identified by flow cytometric assessment combined with genetic analysis. J Allergy Clin Immunol 108(6):1012–1020
24. Conley ME, Brown P, Pickard AR, Buckley RH, Miller DS, Raskind WH, Singer JW, Fialkow PJ (1986) Expression of the gene defect in X-linked agammaglobulinemia. N Engl J Med 315(9):564–567
25. Plebani A, Soresina A, Rondelli R, Amato GM, Azzari C, Cardinale F, Cazzola G, Consolini R, De Mattia D, Dell'Erba G et al (2002) Clinical, immunological, and molecular analysis in a large cohort of patients with X-linked agammaglobulinemia: an Italian multicenter study. Clin Immunol 104(3):221–230
26. Baum C, von Kalle C, Staal FJ, Li Z, Fehse B, Schmidt M, Weerkamp F, Karlsson S, Wagemaker G, Williams DA (2004) Chance or necessity? Insertional mutagenesis in gene therapy and its consequences. Mol Ther 9(1):5–13
27. Howard V, Myers LA, Williams DA, Wheeler G, Turner EV, Cunningham JM, Conley ME (2003) Stem cell transplants for patients with X-linked agammaglobulinemia. Clin Immunol 107(2):98–102
28. Aiuti A (2002) Advances in gene therapy for ADA-deficient SCID. Curr Opin Mol Ther 4(5):515–522
29. Aiuti A, Cattaneo F, Galimberti S, Benninghoff U, Cassani B, Callegaro L, Scaramuzza S, Andolfi G, Mirolo M, Brigida I et al (2009) Gene therapy for immunodeficiency due to adenosine deaminase deficiency. N Engl J Med 360(5):447–458
30. Fischer A, Hacein-Bey-Abina S, Cavazzana-Calvo M (2010) 20 years of gene therapy for SCID. Nat Immunol 11(6):457–460
31. Hacein-Bey-Abina S, von Kalle C, Schmidt M, Le Deist F, Wulffraat N, McIntyre E, Radford I, Villeval JL, Fraser CC, Cavazzana-Calvo M et al (2003) A serious adverse event after

successful gene therapy for X-linked severe combined immunodeficiency. N Engl J Med 348(3):255–256

32. Hacein-Bey-Abina S, Von Kalle C, Schmidt M, McCormack MP, Wulffraat N, Leboulch P, Lim A, Osborne CS, Pawliuk R, Morillon E et al (2003) LMO2-associated clonal T cell proliferation in two patients after gene therapy for SCID-X1. Science 302(5644):415–419
33. Hacein-Bey-Abina S, Garrigue A, Wang GP, Soulier J, Lim A, Morillon E, Clappier E, Caccavelli L, Delabesse E, Beldjord K et al (2008) Insertional oncogenesis in 4 patients after retrovirus-mediated gene therapy of SCID-X1. J Clin Invest 118(9):3132–3142
34. Blank U, Karlsson G, Karlsson S (2008) Signaling pathways governing stem-cell fate. Blood 111(2):492–503
35. Aiuti A, Cassani B, Andolfi G, Mirolo M, Biasco L, Recchia A, Urbinati F, Valacca C, Scaramuzza S, Aker M et al (2007) Multilineage hematopoietic reconstitution without clonal selection in ADA-SCID patients treated with stem cell gene therapy. J Clin Invest 117(8):2233–2240
36. Thrasher AJ, Gaspar HB, Baum C, Modlich U, Schambach A, Candotti F, Otsu M, Sorrentino B, Scobie L, Cameron E et al (2006) Gene therapy: X-SCID transgene leukaemogenicity. Nature 443(7109):E5–E6; discussion E-7
37. Baum C (2007) Insertional mutagenesis in gene therapy and stem cell biology. Curr Opin Hematol 14(4):337–342
38. Staal FJ, Pike-Overzet K, Ng YY, van Dongen JJ (2008) Sola dosis facit venenum. Leukemia in gene therapy trials: a question of vectors, inserts and dosage? Leukemia 22(10):1849–1852
39. Ciuffi A (2008) Mechanisms governing lentivirus integration site selection. Curr Gene Ther 8(6):419–429
40. Yu PW, Tabuchi RS, Kato RM, Astrakhan A, Humblet-Baron S, Kipp K, Chae K, Ellmeier W, Witte ON, Rawlings DJ (2004) Sustained correction of B cell development and function in a murine model of X-linked agammaglobulinemia (XLA) using retroviral-mediated gene transfer. Blood
41. Ellmeier W, Jung S, Sunshine MJ, Hatam F, Xu Y, Baltimore D, Mano H, Littman DR (2000) Severe B cell deficiency in mice lacking the tec kinase family members Tec and Btk. J Exp Med 192(11):1611–1624
42. Kerns HM, Ryu BY, Stirling BV, Sather BD, Astrakhan A, Humblet-Baron S, Liggitt D, Rawlings DJ (2010) B cell-specific lentiviral gene therapy leads to sustained B-cell functional recovery in a murine model of X-linked agammaglobulinemia. Blood 115(11):2146–2155
43. Moreau T, Barlogis V, Bardin F, Nunes JA, Calmels B, Chabannon C, Tonnelle C (2008) Development of an enhanced B-specific lentiviral vector expressing BTK: a tool for gene therapy of XLA. Gene Ther 15(12):942–952
44. Ng YY, Baert MR, Pike-Overzet K, Rodijk M, Brugman MH, Schambach A, Baum C, Hendriks RW, van Dongen JJ, Staal FJ (2010) Correction of B-cell development in Btk-deficient mice using lentiviral vectors with codon-optimized human BTK. Leukemia 24(9):1617–1630
45. Scher I, Ahmed A, Strong DM, Steinberg AD, Paul WE (1975) X-linked B-lymphocyte immune defect in CBA/HN mice. I. Studies of the function and composition of spleen cells. J Exp Med 141(4):788–803
46. Scher I, Steinberg AD, Berning AK, Paul WE (1975) X-linked B-lymphocyte immune defect in CBA/N mice. II. Studies of the mechanisms underlying the immune defect. J Exp Med 142(3):637–650
47. Pike-Overzet K, van der Burg M, Wagemaker G, van Dongen JJ, Staal FJ (2007) New insights and unresolved issues regarding insertional mutagenesis in X-linked SCID gene therapy. Mol Ther 15(11):1910–1916
48. Berglof A, Turunen JJ, Gissberg O, Bestas B, Blomberg KE, Smith CI (2013) Agammaglobulinemia: causative mutations and their implications for novel therapies. Expert Rev Clin Immunol 9(12):1205–1221
49. Bestas B, Moreno PM, Blomberg KE, Mohammad DK, Saleh AF, Sutlu T, Nordin JZ, Guterstam P, Gustafsson MO, Kharazi S et al (2014) Splice-correcting oligonucleotides restore BTK function in X-linked agammaglobulinemia model. J Clin Invest 124(9):4067–4081

50. Hoffman EP, Connor EM (2013) Orphan drug development in muscular dystrophy: update on two large clinical trials of dystrophin rescue therapies. Discov Med 16(89):233–239
51. de Rooij MF, Kuil A, Geest CR, Eldering E, Chang BY, Buggy JJ, Pals ST, Spaargaren M (2012) The clinically active BTK inhibitor PCI-32765 targets B-cell receptor- and chemokine-controlled adhesion and migration in chronic lymphocytic leukemia. Blood 119(11): 2590–2594
52. Harrison C (2012) Trial watch: BTK inhibitor shows positive results in B cell malignancies. Nat Rev Drug Discov 11(2):96
53. Hendriks RW (2011) Drug discovery: new Btk inhibitor holds promise. Nat Chem Biol 7(1):4–5
54. Shinohara M, Chang BY, Buggy JJ, Nagai Y, Kodama T, Asahara H, Takayanagi H (2014) The orally available Btk inhibitor ibrutinib (PCI-32765) protects against osteoclast-mediated bone loss. Bone 60:8–15
55. van den Akker E, van Dijk TB, Schmidt U, Felida L, Beug H, Lowenberg B, von Lindern M (2004) The Btk inhibitor LFM-A13 is a potent inhibitor of Jak2 kinase activity. Biol Chem 385(5):409–413
56. Wodarz D, Garg N, Komarova NL, Benjamini O, Keating MJ, Wierda WG, Kantarjian H, James D, O'Brien S, Burger JA (2014) Kinetics of CLL cells in tissues and blood during therapy with the BTK inhibitor ibrutinib. Blood 123(26):4132–4135
57. Genovese P, Schiroli G, Escobar G, Di Tomaso T, Firrito C, Calabria A, Moi D, Mazzieri R, Bonini C, Holmes MC et al (2014) Targeted genome editing in human repopulating haematopoietic stem cells. Nature 510(7504):235–240

Bruton's Tyrosine Kinase (BTK) Beyond B Lymphocytes: A Protein Kinase with Relevance in Innate Immunity

7

G. Lopez-Herrera, J.L. Maravillas-Montero, J.C. Rodríguez-Alba, and L. Santos-Argumedo

7.1 Introduction

7.1.1 The Tec-Kinase Family and BTK Activation

BTK belongs to a well-studied family of tyrosine kinases known as Tec family kinases (TFKs), which comprises five members: (a) IL-2 inducible T cell kinase (Itk, also identified as Emt or Tsk), (b) Bruton's tyrosine kinase (BTK), (c) tyrosine kinase expressed in hepatocellular carcinoma (Tec), (d) resting lymphocyte kinase (Rlk, also known as Txk), and (e) bone marrow kinase (Bmx). Interestingly, these proteins are differentially expressed by hematopoietic cells, as depicted by previously reported proteomic approaches (Fig. 7.1). Most TFKs are associated with human diseases. Homozygous mutations in *ITK* are associated with a susceptibility to Epstein-Barr virus infection [1], and linkage studies have shown that Tec/Txk locus association with rheumatoid arthritis [2] and mutations of *BTK* are well known to be the cause of X-linked agammaglobulinemia (XLA).

G. Lopez-Herrera
Unidad de Investigación en Inmunodeficiencias, Instituto Nacional de Pediatría-SSa. Cuicuilco, Coyoacán, México City, D.F., México

J.L. Maravillas-Montero
Red de Apoyo a la Investigación, Instituto Nacional de Ciencias Médicas y Nutrición "Salvador Zubirán"-Universidad Nacional Autónoma de México, Tlalpan, México City, D.F., México

J.C. Rodríguez-Alba
Departamento de Biomedicina, Instituto de Ciencias de la Salud, Universidad Veracruzana Campus Xalapa, Xalapa, Veracruz, México

L. Santos-Argumedo (✉)
Departamento de Biomedicina Molecular, Centro de Investigación y de Estudios Avanzados-IPN, Zacatenco, México City, D.F., México
e-mail: lesantos@cinvestav.mx

A. Plebani, V. Lougaris (eds.), *Agammaglobulinemia*, Rare Diseases of the Immune System 4, DOI 10.1007/978-3-319-22714-6_7

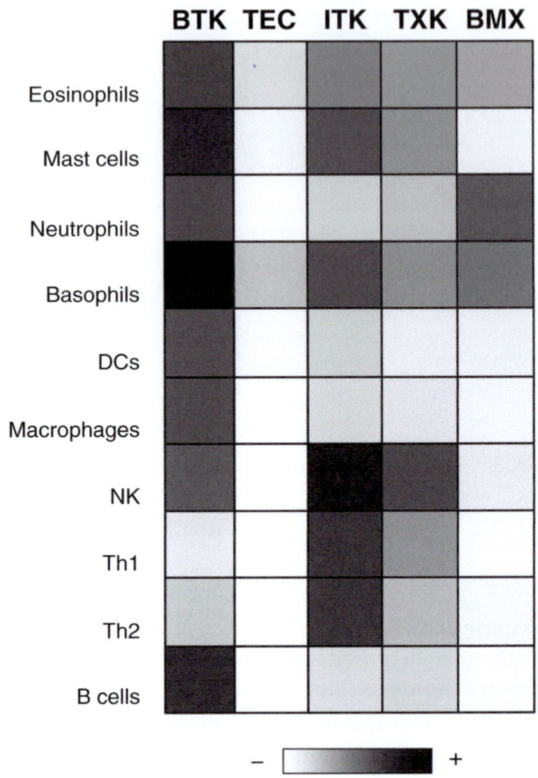

Fig. 7.1 Expression of Tec kinase family members in major human leukocyte populations. The figure represents the level of expression of the five members of the Tec kinase family among different subsets of human leukocytes at the protein level and at the scale provided (The data used to generate this graph were obtained from the ImmGen consortium (www.immgen.org))

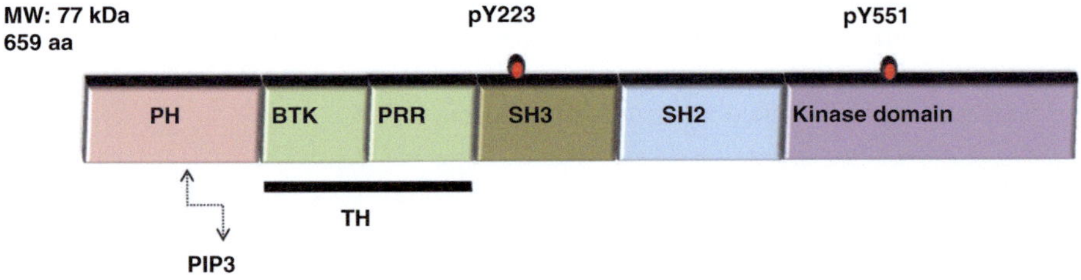

Fig. 7.2 BTK structure. The functional domains of BTK, from the N-terminus: Pleckstrin homology (*PH*); Tec homology (*TH*), formed by the proline-rich region (*PRR*); and the BTK motif, the SH3 and SH2 domains, and the kinase domain. Y223 and Y551 are the tyrosine residues that are phosphorylated for BTK activation. The PH domain is important for membrane localization with PIP3, and the TH domain is important for protein folding and activity regulation. SH3 and SH2 are involved in protein-protein interactions, and finally, the kinase domain is the catalytic domain involved in target phosphorylation

BTK is a 77-kDa protein containing 659 amino acids that can be divided into five different functional domains: the pleckstrin homology (PH) domain is proximal to the N-terminus, followed by the Tec homology (TH), Src homology 3 (SH3), SH2, and SH1 or tyrosine kinase (TK) domains (Fig. 7.2). The PH domain is important for membrane localization of the protein, as it has an affinity for phosphatidylinositol [3–5]-trisphosphate (PIP3). The TH domain contains a proline-rich region (PRR) that is responsible for intramolecular interaction with the SH3 domain, which is important for regulation of BTK activity, as this interaction prevents the phosphorylation of Tyr223, a tyrosine important for mediating full activation of BTK [3]. The SH2 domain of BTK is crucial for cell signaling, as this domain is responsible for interactions with the adaptor protein BLNK (B cell linker protein), which links BTK with Ca^{2+} and MAPK signaling [4]. Additionally, several BTK interacting/regulating proteins have been identified (for a comprehensive review, see [5]). The TK domain is the catalytic domain of BTK; it contains Tyr551, which is phosphorylated by upstream signaling molecules. Once phosphorylated Tyr551, Tyr223 is phosphorylated and BTK becomes fully active [6, 7].

BTK, like most TEC kinases, is preferentially expressed in hematopoietic cells. In particular, BTK is expressed in platelets, erythrocytes, osteoclasts, mast cells, eosinophils, monocytes, macrophages, neutrophils, and in dendritic, NK, and B cells. BTK expression is downregulated in T lymphocytes and plasma cells [8]. This chapter focuses on summarizing the receptors through which BTK is activated in non-B cells and the consequences of BTK deficiency for these cells. Essentially, most of the receptors involved in BTK activation, including the BCR, contain immunotyrosine-based activation motifs (ITAMs), i.e., glycoprotein VI (GPVI) in platelets and RANKL in osteoclasts. Consequently, Src kinases such as Lyn or Fyn are also involved in the initial signaling pathways for BTK activation, leading to activation of phospholipase C (PLC) isoforms and Ca^{2+} release. BTK deficiency in different cell types alters cellular function in various ways, which will be reviewed below. Furthermore, as indicated in some of the cell types reviewed, the function of BTK in several signaling pathways is being studied for its potential use in therapy of various diseases.

7.2 BTK Function in Non-B Cells

7.2.1 BTK in Platelets

BTK is expressed by megakaryocytes and platelets; these cells are activated by collagen through the receptor GPVI, which is associated with the Fcγ receptor (FcγRI). FcγRI contains ITAMs [9]. Activation through GPVI leads to initiation of signaling events (similar to BCR activation), such as Fyn and Lyn activation, which in turn activate SYK, BTK, and PLCγ2 [9] (Fig. 7.3). Activation of platelets by collagen-related peptide (CRP) triggers overproduction of PIP3 in SH2 domain-containing inositol 5-phosphatase (SHIP)-deficient mice, which is correlated with BTK hyperphosphorylation and Ca^{2+} release [10]. The participation of BTK in collagen-activated platelets was confirmed by Oda et al. [11], who also reported that PI3K and SLP-76 participate in the collagen-activated signaling pathway and that in the absence of BTK, human platelets can still be activated by Tec. Similar observations have been reported by Quek et al. [12], who observed that platelets from XLA

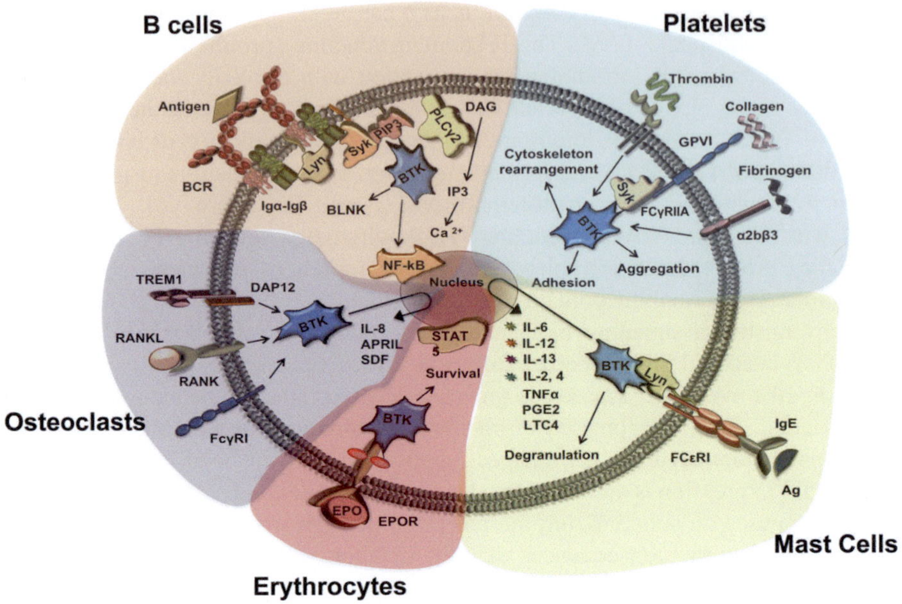

Fig. 7.3 BTK participation in signaling pathways mediated by diverse receptors. Canonical BTK signaling is present in B cells (*Bs*) and involves the participation of the Src kinases, SYK, and Lyn and the subsequent activation of BTK and PLCγ2, which in turn catalyzes Ca^{2+} mobilization. BTK is also involved in the signaling pathway mediated by thrombin and collagen receptors (*GPVI*) in platelets (*Pl*). The activation of these pathways induces platelet aggregation, dense granule secretion, and cytoskeleton rearrangement. BTK participates in FcεRI-dependent mast cell activation, and the interaction of this receptor leads to the induction of cytokines and leukotriene C4 (*LTC4*) production. Participation of BTK with erythropoietin receptor favors erythrocyte (*ER*) survival and the response to SDF by c-Kit. Finally, BTK participation in osteoclasts (*OS*) involves the signaling pathway mediated by RANK, TREM-1, and FcRg, which are involved in the induction of growth factors (APRIL, IL-8, and SDF-1) and in favoring osteoclastogenesis and bone resorption

patients showed reduced aggregation and dense granule secretion, as shown by deficient 5-hydroxytryptamine secretion (5HT) in response to collagen. The reduced but not absent responses to collagen or CRP may be explained by the participation of Tec, another TFK, in the GPVI-mediated signaling pathway. Tec is also expressed in platelets; *Btk*$^{-/-}$/*Tec*$^{-/-}$ mice show a dramatic reduction of PLCγ2 activation compared to *Btk*$^{-/-}$ or *Tec*$^{-/-}$ mice [13]. GPVI-dependent platelet shape changes and aggregation are reduced in *Btk*$^{-/-}$ or *Tec*$^{-/-}$ mice and severely affected in *Btk*$^{-/-}$/*Tec*$^{-/-}$ mice, indicating an important role for both kinases in proper platelet function. However, BTK plays a more prominent role than Tec, as platelet function is more strongly affected in *Btk*$^{-/-}$ mice, as shown by the abolishment of α-granule secretion in *Btk*$^{-/-}$ but not in *Tec*$^{-/-}$ mice [13].

BTK phosphorylation also occurs after stimulation with thrombin, but platelet aggregation and dense granule secretion are not altered [12]. In human platelets, Mukhopadhyay et al. [14] suggest that BTK interacts with the actin cytoskeleton through FcγRIIA (Fig. 7.3) after stimulation with thrombin. However, these data are based on analysis of the Triton X100-insoluble platelet fraction, and neither colocalization nor immunoprecipitation assays have been performed [14].

A second receptor that has been suggested to activate BTK is the integrin $\alpha_{IIb}\beta_3$, which mediates adhesion to fibrinogen and induces platelet aggregation [15] (Fig. 7.3). Ligation of $\alpha_{IIb}\beta_3$ using fibrinogen-coated surfaces induces phosphorylation of SYK, SLP-76, and BTK and in consequence of PLCγ2; this signaling pathway has been suggested to reduce actin polymerization-dependent phenomena, such as formation of filopodia and lamellipodia in platelets from *Plcg2*$^{-/-}$ mice, after exposure to fibrinogen [15].

Platelets, apart from their role in plug formation in damaged blood vessels, also play an important role in inflammation, as they can promote leukocyte infiltration through interaction with leukocytes and endothelial cells [16]. Indeed, depletion of platelets reduces the accumulation of virus-specific cytotoxic T cells [17]. Additionally, infection with lymphocytic choriomeningitis virus in mice, coupled with platelet depletion, induces lethal hemorrhagic anemia [18]. No reports have been published indicating that platelet dysfunction alters either plug formation or leukocyte recruitment in patients with XLA; however, it has been observed that blocking BTK in platelets reduces the formation of platelet microparticles (small membrane-coated vesicles that are released when platelets are activated) [19]. The release of these particles has been linked to the pathogenesis of rheumatoid arthritis, due to their interaction with synoviocytes, which induces these cells to produce pro-inflammatory cytokines [20]. Therefore, BTK inhibition plays a negative role in the formation of these particles and, in consequence, reduces induction of pro-inflammatory cytokines such as IL-6 and IL-8 by synoviocytes [19].

7.2.2 BTK in Erythrocytes

BTK was found to be expressed in the human erythroleukemia cell line K-562 [8] and in the progenitors of chicken erythrocytes and chicken oncogene-transformed erythroblasts [21]. Few studies have been conducted on the role of BTK in erythroid signaling pathways; however, interestingly, Schmidt et al. [22] reported that BTK is involved in the signaling mediated by the erythropoietin (Epo) and stem cell factor (SCF or Kit-ligand) receptors (EpoR and c-Kit or CD117, respectively). Both cytokines are crucial for the development of erythroid progenitors [22]. In this paper, the authors demonstrated that BTK is phosphorylated after Epo and SCF stimulation in primary erythroid progenitors and that BTK activation is important for the proper response to Epo and for SCF-mediated survival [22]. They also demonstrated that BTK participates in the EpoR-mediated signaling pathway, which includes phosphorylation of PLCγ1, JAK2, and STAT5 (Fig. 7.3). Although the role of BTK in c-Kit signaling is unclear, the authors suggested that BTK interacts with the dead receptor TRAILR1 and that such interactions are dependent on SCF [22]. In fact, it has been proposed that c-Kit activation induces LYN activation, which in turn leads to BTK-TRAILR1 interaction, and that such interactions inhibit TRAILR1-mediated apoptosis [23].

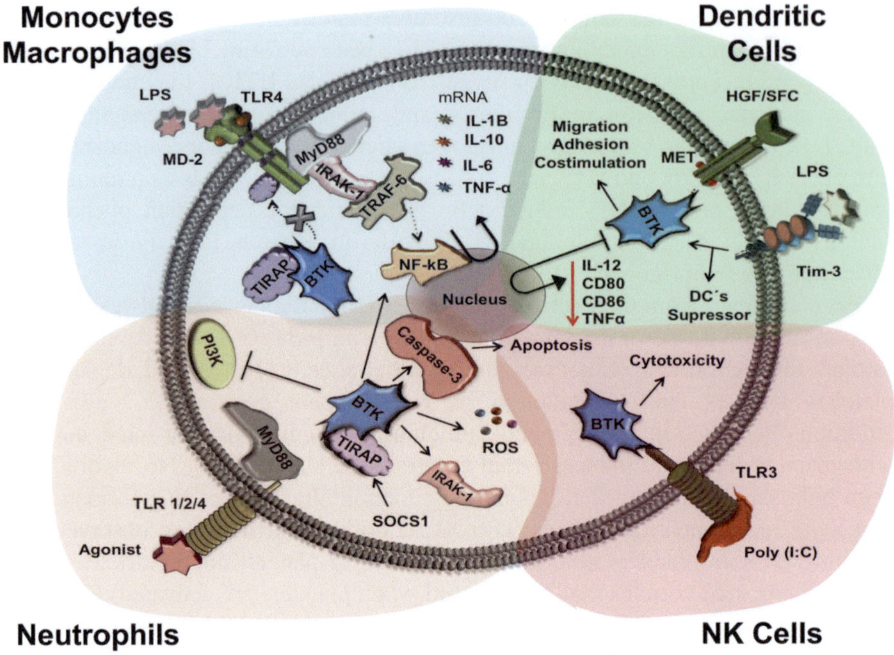

Fig. 7.4 TLR-dependent activation of BTK in monocytes, dendritic cells, NK cells, and neutrophils. BTK acts as an inhibitor of TLR-dependent activation, by interacting with TIRAP in monocytes. This interaction favors TIRAP degradation, negatively regulating NF-κB activation by TLR agonists. The effect of BTK in TLR-mediated activation in different cell types ranges from negative regulation of pro-inflammatory cytokine production (monocytes), the induction of cytokines, and activation molecules by dendritic cells, favoring of NK cell-mediated cytotoxicity, to negative regulation of the production of reactive oxygen species (ROS) by neutrophils

7.2.3 BTK in NK Cells

The function of BTK in NK cells has been poorly explored; Futatani et al. [24] reported that BTK is not expressed in human NK cells. However, Bao et al. [25] recently reported that BTK is expressed in murine resting NK cells and that BTK expression is upregulated after activation and during maturation. Additionally, the authors reported that BTK is phosphorylated after stimulation with the TLR3 ligand poly (I:C) and that stimulation of NK cells from $BTK^{-/-}$ mice results in decreased cytotoxicity activity and reduced numbers of NK cells. Finally, NK cells from XLA patients also showed defects in cytotoxicity, although XLA patients and healthy donors show comparable levels of NK cells [25] (Fig. 7.4).

7.2.4 BTK in Osteoclasts

Osteoclasts, the cells responsible for bone resorption, are cells of myeloid origin that also express BTK [26]. Differentiation in vitro, via M-CSF and RANKL, of osteoclasts from bone marrow-derived monocyte/macrophage (BMM) precursors is severely impaired in *Btk*$^{-/-}$ mice and almost absent in *Btk*$^{-/-}$/Tec$^{-/-}$ mice. The expression of NFATc1, the key nuclear factor for osteoclast diferentiation, is reduced in *Btk*$^{-/-}$/*Tec*$^{-/-}$ mice. Both BTK and Tec become phosphorylated after RANKL stimulation of BMMs [26]. The role of BTK in osteoclastogenesis seems to require a functional SH1 domain and is independent of the PH domain, as reconstitution with mutated R525Q BTK in *Tec*$^{-/-}$/*Btk*$^{-/-}$ osteoclasts blocks in vitro osteoclast formation, while R28C does not [26]. Similar observations were presented by Lee et al. [27] using *xid* mice, which carry a mutation in the R28C residue; *xid* mice also show defects in osteoclast differentiation accompanied by reduced resorption clusters or pit formation, reduced cell-cell fusion, and defective formation of multinucleated osteoclasts. Although *xid* mice show in vitro defects of osteoclast differentiation, *Btk*$^{-/-}$-deficient mice do not exhibit increased bone density; an increase in bone density has been detected only in *Tec*$^{-/-}$/*Btk*$^{-/-}$ mice [26]. Patients with XLA do not exhibit changes in bone density, but osteoclast function is affected; in this case, osteoclasts from XLA patients show defects in pit formation due to defects in actin polymerization, but in vitro differentiation of osteoclasts from peripheral blood CD14+ cells is increased. However, in vitro cultures of XLA osteoclasts with XLA serum, which show increased levels of TNF-α, IL-6, and IL-1β, show restoration of pit formation, indicating that XLA patients' normal bone density might be due to the osteoclastogenic activity of XLA serum and the activity of Tec [28]. The signaling pathway in which BTK appears to participate in osteoclasts is mediated by RANK, FcγR, and DAP12 [29, 30] (Fig. 7.3). Fc receptor common γ–subunit (FcRγ) and DAP12 associate with several membrane receptors; after activation, they induce Ca^{2+} release through PLCγ2 activation (for a review, see [31]). In addition, TREM-1/DAP12 signaling has been reported to operate through BTK in the U-937 myeloid cell line [32].

Several studies on the use of BTK inhibitors in autoimmune diseases and cancer have been reported recently; for example, the use of PCI-32765 in an inducible model of rheumatoid arthritis significantly reduces clinical symptoms, such as joint inflammation and cartilage and bone damage, in comparison to dexamethasone [33]. The effects of PCI-32765 have also been studied in multiple myeloma, in which the pathogenesis of cancer cells depends on the bone marrow microenvironment and interactions between osteoclasts, osteoblasts, and bone marrow stromal cells and the growth factors secreted by these cells [34]. The use of PCI-32765 inhibits osteoclast differentiation and suppresses, in a dose-dependent manner, osteoclast-dependent proliferation and survival of myeloma cells. It also has direct effects on myeloma cells, as these cells also express BTK [35]. Moreover, this BTK inhibitor also inhibits the secretion of growth factors by osteoclasts such as IL-8, CXCL12, and APRIL, among others, indicating that this inhibitor may be useful in treating multiple myeloma [35]. Further data provided by Bam et al. [36] indicate that LFM-A13, a BTK inhibitor, suppresses the migration of osteoclasts and myeloma cells in response to CXCL12, indicating that osteoclasts also play an important role in the pathogenesis of multiple myeloma [36] and that the BTK inhibitors that are under investigation for treatment of lymphomas can potentially be used in the treatment of multiple myeloma [37]. In fact, PCI-32765 inhibits proliferation and induces cytotoxicity in vitro of chemotherapy-treated myeloid leukemia cells [38]. For a comprehensive recent review, see Seng-Lai Tan et al. [39].

7.2.5 BTK in Mast Cells

Activation of mast cells is mainly driven through the interaction of the high-affinity receptor FcεRI with IgE and antigen binding [40] (Fig. 7.3). The role of FcεRI-activated BTK has been studied in mast cells. Yamada et al. [41] initially reported that mast cells express BTK. Recently, Hata et al. [42] showed that after stimulation of bone marrow-derived mast cells (BMMCs) from *xid* mice with dinitrophenyl-IgE conjugates, degranulation and production of histamine and cytokines (including TNFα, IL-2, and IL-6) are deficient, compared to wild-type cells. The results show a reduction in the severity of induced passive cutaneous anaphylaxis in *xid* mice [42]. Additional information regarding the role of BTK in mast cell activation after FcεRI cross-linking indicates an important role for Lyn, as $Lyn^{-/-}/Btk^{-/-}$ mice showed a dramatic reduction in histamine and leukotriene release and decreased phosphorylation of PLCγ1, PLCγ2, and some isoforms of PKC [43]. Kuehn et al. [44] reported similar results for human mast cells, including a significant reduction in the production of reactive oxygen species (ROS), prostaglandin E2 (PGE2), and leukotriene C4 (LTC4), in a FcεRI-dependent manner.

BTK also seems to be important for mast cell migration, as $Btk^{-/-}$ IgE-sensitized BMMCs show reduced migration in response to CXCL12, PGE2, or adenosine, suggesting a synergistic effect between FcεRI and PGE2 in BTK and PLCγ1 phosphorylation. Additionally, a synergistic effect in response to FcεRI and PGE2 was also observed in the activation of Rac1 and the induction of actin polymerization [45]. In addition to BTK, Itk and Tec are also involved in FcεRI-mediated signaling; in the absence of both of these TFKs, defective LTC4 production and cytokine secretion (such as IL-4, GM-CSF, TNFα, and IL-13) are observed in $Tec^{-/-}$ mast cells [46], or a dramatic reduction in IL-2, IL-3, IL-4, and GM-CSF is observed in $Btk^{-/-}/Itk^{-/-}$ but not in $Itk^{-/-}$ or $Btk^{-/-}$-activated mast cells [47].

Further studies have reported a possible role of BTK in mast cell activation through c-Kit or toll-like receptors. Iwaki et al. [48] have reported that stimulation with SCF induces phosphorylation of BTK that is potentiated with SCF and FcεRI; both stimuli also potentiated mast cell degranulation, cytokine production (TNFα, IL-6, and IL-13), and PLCγ1 phosphorylation. Finally, BTK does not seem to play a role via TLR activation in mast cells, as comparable or slight increases in cytokine production are detected in response to ligands for TLR4, TLR2/6, or TLR2/1 [49]. The use of BTK, SYK, and/or Src family kinase inhibitors in the treatment of allergic disease has also been explored, and these have been demonstrated to be effective for preventing mast cell activation at the cellular level and for suppression of pulmonary inflammation in murine models of asthma, indicating their potential for use in the treatment of allergic disease (for a review, see [50]).

7.2.6 BTK in Dendritic Cells

Studies of the role of BTK in dendritic cells (DCs) are typically performed in DCs differentiated from peripheral blood monocytes (mDCs) or bone marrow (BMDCs) from humans and mice, respectively. Important differences have been observed when mDCs or BMDCs are used; mDCs obtained from XLA patients can induce activation and expression of costimulatory molecules by T cells, similar to that observed in WT mice [51, 52]. However, BMDCs obtained from *Btk*$^{-/-}$ mice show reduced secretion of IL-10 in response to LPS, a defect that has been proposed as a mechanism to increase the proliferative response and cytokine secretion (in particular IL-4, IFN γ, IL-5, and IL-13) by T cells [53, 54] (Fig. 7.4). Additionally, Shingal et al. [54] proposed that BTK signaling is mediated by the scatter factor (SF)/hepatocyte growth factor receptor, c-MET, which has been demonstrated to be expressed by DCs and to influence adhesion and migration of these cells [55] (Fig. 7.4).

In addition, mDCs obtained from XLA patients showed activation defects in response to TLR-9 agonists, because they failed to express cytokines (IL-6, IL-12p70, and TNFα) and activation markers (CD80, CD86, HLA-DR, and CD83); mDCs from XLA patients also failed to upregulate STAT1/3 transcripts in response to TLR-9 agonists [56].

CD303 is another receptor that has been reported to activate BTK and various signaling proteins (LYN, SYK, BLNK, and PLCγ2) and to interact with FcγR. CD303 cross-linking has been associated with reduced phosphorylation of IκB and inhibition of IFN-I gene transcription in CpG-stimulated PDCs [57]. T cell Ig and mucin protein 3 (TIM-3), a receptor that recognizes and mediates phagocytosis of apoptotic cells, has also been reported to induce BTK activation in DCs [58]. Cross-linking of TIM-3 in BMDCs induced BTK and c-Src phosphorylation and the suppression of LPS-activated DCs. In particular, it downregulates the secretion of IL-12 and TNFα and the expression of CD40, CD80, and CD86, indicating that BTK plays a regulatory role in the activation of these cells in the TIM-3 pathway [59].

7.2.7 BTK in Neutrophils

The role of BTK in neutrophils is of interest, and the discovery that BTK is able to interact with molecules involved in TLR signaling, such as IRAK-1, MyD88, and TIRAP in monocyte-derived cell lines, makes it clear that BTK participates in functions of the innate immune system functions [60, 61] (Fig. 7.4). Additionally, it has been demonstrated that BTK interacts with and phosphorylates TIRAP, promoting its polyubiquitination and degradation by the suppressor of cytokine signaling-1 (SOCS-1), causing increased activation of NF-κB [62] (Fig. 7.4). The role of BTK in neutrophils obtained from *Btk*$^{-/-}$ mice is demonstrated by the alterations observed in granulopoiesis and granulocyte function (reduced expression of granulocyte transcripts for myeloperoxidase, elastase, and lactoferrin) [63].

Recent publications have focused on the importance of BTK in regulating oxidative burst in neutrophils, and contrasting results have been reported. BTK-deficient neutrophils obtained from XLA patients show normal production of reactive oxygen species (ROS) in response to TLR4 and TLR7/8 agonists [64]; however, another report showed that XLA neutrophils stimulated with TLR-4 or TLR1/2 agonists showed exacerbated ROS production and were prone to apoptosis due to increased levels of caspase-3. An association with increased levels of protein tyrosine kinases activity has also been observed (in particular with Fak, Lyn, Src, SYK, Vav, and PI3K), and PI3K-p85 was found to be associated with TIRAP and to be constitutively phosphorylated in XLA neutrophils, suggesting that the main function of BTK in human neutrophils consists of regulating the activation of PI3K by TIRAP [65].

BTK has also been implicated in FcγRIIa/TLR4 cross talk, as demonstrated by an increased response to immune complexes in LPS-pretreated cells. Krupa et al. [66] showed that LPS pretreatment increases the expression of FcγRIIa, shortening the distance between these two receptors, and that when FcγRIIa is triggered by immune complexes, TLR4 becomes activated. The role of Fc receptors in inflammation, antibody-dependent cytotoxicity, and phagocytosis is so important that many efforts have been made to regulate their undesirable effects. BTK has been considered as a possible target, and efforts have been made to develop specific inhibitors for treatment of diseases, such as rheumatoid arthritis, where immune complexes are involved, with promising results, at least in animal models [67, 68].

BTK may also play a role in neutrophil adhesion and migration, as neutrophils from *Btk*$^{-/-}$ mice show altered rolling velocity after E-selectin engagement, which also induced BTK, PLCγ2, and PI3K phosphorylation [69]. Additionally, chemotaxis is reduced in response to fMLP in human neutrophils treated with LFM-A13 [70, 71]. A comprehensive review of this topic has recently been published by Block and Zarbock [72].

Recent reports have focused on the inhibition of BTK, to determine whether it is useful in the treatment of some diseases, for example, the inhibition of BTK expression by siRNA-restored lung homeostasis in a model of murine acute lung injury induced by injection with immune complexes [73].

7.2.8 BTK in Monocytes/Macrophages

The expression of BTK in monocytes/macrophages suggests that it plays a functional role in this cell lineage. Numerous studies have indicated a role for Tec family kinases (mostly BTK) in TLR signaling pathways. Initially, immunoprecipitation experiments performed with the human promonocytic cell line THP1 demonstrated a robust interaction between BTK and several members of the TLR signaling pathway such as IRAK4, Myd88, and TIRAP, but not TRAF6 [60] (Fig. 7.4). Furthermore, the TLR4 ligand LPS induces activation of Tec and BTK kinases in human monocytes [60, 74], and the TIRAP adaptor protein is phosphorylated by BTK through TLR2- and TLR4-induced signaling [75, 76] (Fig. 7.4).

The precise role of BTK in LPS-induced signaling that leads to transcription of TNFα is complex. BTK-deficient human monocytes show reduced LPS-induced TNFα production after 18 h of stimulation [74]. Mature macrophages from *xid mice* (lacking a functional BTK protein) show reduced LPS-induced TNFα production after 48 h of stimulation [77], whereas BTK-deficient human monocytes stimulated for 4 h with LPS showed no alteration in TNFα production [78]. These differences may depend on the period of stimulation, and longer periods may result in secondary signaling effects.

Approximately 15 years ago, several studies showed that *xid* mice are resistant to the induction of some inflammatory diseases [42, 79]. These experiments suggested that the naturally occurring BTK mutation found in these mice was likely to play a role in inflammation. In these experiments, endotoxin-treated *xid* macrophages showed impaired production of reactive oxygen intermediates (ROIs) and nitric oxide (NO) [76]. TLR activation was later identified as the initiating event, and further attention has since been paid to the activity of specific TLR pathways in *xid* as well as BTK-knockout mice. However, these studies have reported contradictory results, suggesting both hypo- and hyper-responsiveness to TLR ligands in the absence of BTK. Studies of *xid* macrophages demonstrated increases in pro-inflammatory IL-6 in response to TLR activation in conjunction with decreased IL-10 [80]. BTK is phosphorylated following activation of both extracellular TLRs and endosomal TLRs, and yeast-2-hybrid studies have suggested a direct interaction of BTK with the cytoplasmic toll/interleukin-l receptor (TIR) domains of TLRs 4, 6, 8, and 9 [60]. LPS-induced activation of PI3K in the plasma membrane establishes a PIP3 gradient thought the recruitment of BTK to activated TLRs and associated Src kinases [7, 81, 82]. BTK appears to participate in further molecular associations in the proximal TLR4 signalosome; it can be immunoprecipitated with TIRAP, MyD88, and IRAK1 after treatment of cells with LPS [60, 83]. Several recent studies have further suggested that BTK is responsible for phosphorylation of TIRAP, leading to its degradation and to disruption of TLR4 signaling [62, 75]. This would suggest that BTK is a vital regulator of TLR-mediated pro-inflammatory signaling in monocytes/macrophages.

BTK has been implicated in TLR3 phosphorylation, which is known to be important in the development of innate immune responses to viruses. BTK-deficient macrophages are unable to produce IFN-β upon stimulation with poly (I:C) and were less able to clear a dengue virus infection than unstimulated macrophages [84]. Thus, BTK is also important in the initiation of TLR3 signaling, similar to its interactions with other TLRs.

The role of BTK in the signaling pathways of several TLRs indicates that the molecule may participate in the polarization of macrophages into M1 or M2 phenotypes. As described above, BTK interacts with TLR2, TLR3, TLR4, and TLR7, mediating their phosphorylation and in consequence their signal transduction (Fig. 7.4). LPS induces cytokine production in macrophages, driving their differentiation into an M1 phenotype. Thus, M1 differentiation was explored in BTK-deficient mice, and impaired polarization to M1, with an increased number of M2 macrophages, was observed. At the molecular level, macrophages from BTK-deficient mice showed decreased STAT1 and increased levels of STAT6 phosphorylation. These results may indicate exacerbation of the inflammatory allergic response, because a higher number of M2 macrophages were recruited upon in vivo stimulation with *Schistosoma mansoni* [85].

In a more complex setting, BTK becomes activated in macrophages infected with *Listeria monocytogenes*, and BMMs derived from BTK-deficient mice infected with *Listeria* produce larger amounts of TNF-α, IL-6, IL-12p70, and IL-10 but similar amounts of IFN-I and NO. Thus, BTK plays a regulatory role under these conditions, limiting the production of inflammatory cytokines [86]. These results are concordant with clinical observations reported by Gonzalez-Serrano et al. [87], who observed increased pro-inflammatory cytokine production in XLA patients.

Stimulation with LPS plus IFN-γ induces apoptosis of macrophages, and BTK deficiency results in an enhanced susceptibility to cell death. This enhanced susceptibility is specific to these two stimuli, and Fas/CD95 engagement does not produce these differences. Thus, BTK also links TLR4 and IFN-γ signaling [88]. However, macrophages lacking BTK are less efficient at removing apoptotic cells; the mechanism underlying these processes involves the chaperone calreticulin (CRT). Byrne et al. demonstrated that CTR is a new target of BTK and provided a new, unexpected role for this tyrosine kinase [89].

In general, Tec family kinases are involved in PLCγ2 activation and Ca^{2+} regulation and are important in many signaling events, such as arachidonic acid release by macrophages. However, the mode of BTK action remains unclear. BTK activation results in increased inositol (1, 4, 5)-trisphosphate (IP3) and depletion of Ca^{2+} stores [90]. Furthermore, BTK can affect the synthesis of phosphatidylinositol (4,5)-bisphosphate or PIP2, a substrate for both PLCγ2 and PI3K, thereby influencing PLCγ2 activation and Ca^{2+} regulation [77].

Conclusions

In summary, several novel, exciting roles of BTK have emerged during the last 20 years. Despite its critical function in BCR signal transduction, BTK has been shown to participate in the signal transduction of many receptors, in manners similar to or different from its role in BCR signaling, involving regulation of fundamental functions such as inflammation and apoptosis. These new roles make BTK an attractive target for treating not only malignant conditions but also inflammatory diseases, where it may offer new tools to control various signaling pathways in which BTK is involved, using small molecules.

Acknowledgments This work was supported by Instituto de Ciencia y Tecnología del Distrito Federal (ICyTDF) (Grant PICSA12-157) and Consejo Nacional de Ciencia y Tecnología (CONACyT) (Grant Salud-202111 and CB-154472).

References

1. Huck K, Feyen O, Niehues T, Ruschendorf F, Hubner N, Laws HJ et al (2009) Girls homozygous for an IL-2-inducible T cell kinase mutation that leads to protein deficiency develop fatal EBV-associated lymphoproliferation. J Clin Invest 119(5):1350–1358
2. Freudenberg J, Lee AT, Siminovitch KA, Amos CI, Ballard D, Li W et al (2010) Locus category based analysis of a large genome-wide association study of rheumatoid arthritis. Hum Mol Genet 19(19):3863–3872

3. Lindvall JM, Blomberg KE, Valiaho J, Vargas L, Heinonen JE, Berglof A et al (2005) Bruton's tyrosine kinase: cell biology, sequence conservation, mutation spectrum, siRNA modifications, and expression profiling. Immunol Rev 203:200–215
4. Chiu CW, Dalton M, Ishiai M, Kurosaki T, Chan AC (2002) BLNK: molecular scaffolding through 'cis'-mediated organization of signaling proteins. EMBO J 21(23):6461–6472
5. Mohamed AJ, Yu L, Backesjo CM, Vargas L, Faryal R, Aints A et al (2009) Bruton's tyrosine kinase (Btk): function, regulation, and transformation with special emphasis on the PH domain. Immunol Rev 228(1):58–73
6. Park H, Wahl MI, Afar DE, Turck CW, Rawlings DJ, Tam C et al (1996) Regulation of Btk function by a major autophosphorylation site within the SH3 domain. Immunity 4(5):515–525
7. Rawlings DJ, Scharenberg AM, Park H, Wahl MI, Lin S, Kato RM et al (1996) Activation of BTK by a phosphorylation mechanism initiated by SRC family kinases. Science 271(5250):822–825
8. Smith CI, Baskin B, Humire-Greiff P, Zhou JN, Olsson PG, Maniar HS et al (1994) Expression of Bruton's agammaglobulinemia tyrosine kinase gene, BTK, is selectively down-regulated in T lymphocytes and plasma cells. J Immunol 152(2):557–565
9. Bergmeier W, Stefanini L (2013) Platelet ITAM signaling. Curr Opin Hematol 20(5):445–450
10. Pasquet JM, Quek L, Stevens C, Bobe R, Huber M, Duronio V et al (2000) Phosphatidylinositol 3,4,5-trisphosphate regulates Ca(2+) entry via btk in platelets and megakaryocytes without increasing phospholipase C activity. EMBO J 19(12):2793–2802
11. Oda A, Ikeda Y, Ochs HD, Druker BJ, Ozaki K, Handa M et al (2000) Rapid tyrosine phosphorylation and activation of Bruton's tyrosine/Tec kinases in platelets induced by collagen binding or CD32 cross-linking. Blood 95(5):1663–1670
12. Quek LS, Bolen J, Watson SP (1998) A role for Bruton's tyrosine kinase (Btk) in platelet activation by collagen. Curr Biol 8(20):1137–1140
13. Atkinson BT, Ellmeier W, Watson SP (2003) Tec regulates platelet activation by GPVI in the absence of Btk. Blood 102(10):3592–3599
14. Mukhopadhyay S, Ramars AS, Dash D (2001) Bruton's tyrosine kinase associates with the actin-based cytoskeleton in activated platelets. J Cell Biochem 81(4):659–665
15. Wonerow P, Pearce AC, Vaux DJ, Watson SP (2003) A critical role for phospholipase Cgamma2 in alphaIIbbeta3-mediated platelet spreading. J Biol Chem 278(39):37520–37529
16. Gibbins JM (2004) Platelet adhesion signalling and the regulation of thrombus formation. J Cell Sci 117(Pt 16):3415–3425
17. Iannacone M, Sitia G, Isogawa M, Marchese P, Castro MG, Lowenstein PR et al (2005) Platelets mediate cytotoxic T lymphocyte-induced liver damage. Nat Med 11(11):1167–1169
18. Iannacone M, Sitia G, Isogawa M, Whitmire JK, Marchese P, Chisari FV et al (2008) Platelets prevent IFN-alpha/beta-induced lethal hemorrhage promoting CTL-dependent clearance of lymphocytic choriomeningitis virus. Proc Natl Acad Sci U S A 105(2):629–634
19. Hsu J, Gu Y, Tan SL, Narula S, DeMartino JA, Liao C (2013) Bruton's Tyrosine Kinase mediates platelet receptor-induced generation of microparticles: a potential mechanism for amplification of inflammatory responses in rheumatoid arthritis synovial joints. Immunol Lett 150(1-2):97–104
20. Knijff-Dutmer EA, Koerts J, Nieuwland R, Kalsbeek-Batenburg EM, van de Laar MA (2002) Elevated levels of platelet microparticles are associated with disease activity in rheumatoid arthritis. Arthritis Rheum 46(6):1498–1503
21. Robinson D, Chen HC, Li D, Yustein JT, He F, Lin WC et al (1998) Tyrosine kinase expression profiles of chicken erythro-progenitor cells and oncogene-transformed erythroblasts. J Biomed Sci 5(2):93–100
22. Schmidt U, van den Akker E, Parren-van Amelsvoort M, Litos G, de Bruijn M, Gutierrez L et al (2004) Btk is required for an efficient response to erythropoietin and for SCF-controlled protection against TRAIL in erythroid progenitors. J Exp Med 199(6):785–795
23. von Lindern M, Schmidt U, Beug H (2004) Control of erythropoiesis by erythropoietin and stem cell factor: a novel role for Bruton's tyrosine kinase. Cell Cycle 3(7):876–879

24. Futatani T, Miyawaki T, Tsukada S, Hashimoto S, Kunikata T, Arai S et al (1998) Deficient expression of Bruton's tyrosine kinase in monocytes from X-linked agammaglobulinemia as evaluated by a flow cytometric analysis and its clinical application to carrier detection. Blood 91(2):595–602
25. Bao Y, Zheng J, Han C, Jin J, Han H, Liu Y et al (2012) Tyrosine kinase Btk is required for NK cell activation. J Biol Chem 287(28):23769–23778
26. Shinohara M, Koga T, Okamoto K, Sakaguchi S, Arai K, Yasuda H et al (2008) Tyrosine kinases Btk and Tec regulate osteoclast differentiation by linking RANK and ITAM signals. Cell 132(5):794–806
27. Lee SH, Kim T, Jeong D, Kim N, Choi Y (2008) The tec family tyrosine kinase Btk regulates RANKL-induced osteoclast maturation. J Biol Chem 283(17):11526–11534
28. Danks L, Workman S, Webster D, Horwood NJ (2011) Elevated cytokine production restores bone resorption by human Btk-deficient osteoclasts. J Bone Miner Res 26(1):182–192
29. Koga T, Inui M, Inoue K, Kim S, Suematsu A, Kobayashi E et al (2004) Costimulatory signals mediated by the ITAM motif cooperate with RANKL for bone homeostasis. Nature 428(6984):758–763
30. Mocsai A, Humphrey MB, Van Ziffle JA, Hu Y, Burghardt A, Spusta SC et al (2004) The immunomodulatory adapter proteins DAP12 and Fc receptor gamma-chain (FcRgamma) regulate development of functional osteoclasts through the Syk tyrosine kinase. Proc Natl Acad Sci U S A 101(16):6158–6163
31. Lanier LL (2009) DAP10- and DAP12-associated receptors in innate immunity. Immunol Rev 227(1):150–160
32. Ormsby T, Schlecker E, Ferdin J, Tessarz AS, Angelisova P, Koprulu AD et al (2011) Btk is a positive regulator in the TREM-1/DAP12 signaling pathway. Blood 118(4):936–945
33. Chang BY, Huang MM, Francesco M, Chen J, Sokolove J, Magadala P et al (2011) The Bruton tyrosine kinase inhibitor PCI-32765 ameliorates autoimmune arthritis by inhibition of multiple effector cells. Arthritis Res Ther 13(4):R115
34. Raje N, Roodman GD (2011) Advances in the biology and treatment of bone disease in multiple myeloma. Clin Cancer Res 17(6):1278–1286
35. Tai YT, Chang BY, Kong SY, Fulciniti M, Yang G, Calle Y et al (2012) Bruton tyrosine kinase inhibition is a novel therapeutic strategy targeting tumor in the bone marrow microenvironment in multiple myeloma. Blood 120(9):1877–1887
36. Bam R, Ling W, Khan S, Pennisi A, Venkateshaiah SU, Li X et al (2013) Role of Bruton's tyrosine kinase in myeloma cell migration and induction of bone disease. Am J Hematol 88(6):463–471
37. Uckun FM, Qazi S (2010) Bruton's tyrosine kinase as a molecular target in treatment of leukemias and lymphomas as well as inflammatory disorders and autoimmunity. Expert Opin Ther Pat 20(11):1457–1470
38. Rushworth SA, Murray MY, Zaitseva L, Bowles KM, MacEwan DJ (2014) Identification of Bruton's tyrosine kinase as a therapeutic target in acute myeloid leukemia. Blood 123(8):1229–1238
39. Tan SL, Liao C, Lucas MC, Stevenson C, DeMartino JA (2013) Targeting the SYK-BTK axis for the treatment of immunological and hematological disorders: recent progress and therapeutic perspectives. Pharmacol Ther 138(2):294–309
40. Turner H, Kinet JP (1999) Signalling through the high-affinity IgE receptor Fc epsilonRI. Nature 402(6760 Suppl):B24–B30
41. Yamada N, Kawakami Y, Kimura H, Fukamachi H, Baier G, Altman A et al (1993) Structure and expression of novel protein-tyrosine kinases, Emb and Emt, in hematopoietic cells. Biochem Biophys Res Commun 192(1):231–240
42. Hata D, Kawakami Y, Inagaki N, Lantz CS, Kitamura T, Khan WN et al (1998) Involvement of Bruton's tyrosine kinase in FcepsilonRI-dependent mast cell degranulation and cytokine production. J Exp Med 187(8):1235–1247
43. Kawakami Y, Kitaura J, Satterthwaite AB, Kato RM, Asai K, Hartman SE et al (2000) Redundant and opposing functions of two tyrosine kinases, Btk and Lyn, in mast cell activation. J Immunol 165(3):1210–1219
44. Kuehn HS, Swindle EJ, Kim MS, Beaven MA, Metcalfe DD, Gilfillan AM (2008) The phosphoinositide 3-kinase-dependent activation of Btk is required for optimal eicosanoid production and generation of reactive oxygen species in antigen-stimulated mast cells. J Immunol 181(11):7706–7712

45. Kuehn HS, Radinger M, Brown JM, Ali K, Vanhaesebroeck B, Beaven MA et al (2010) Btk-dependent Rac activation and actin rearrangement following FcepsilonRI aggregation promotes enhanced chemotactic responses of mast cells. J Cell Sci 123(Pt 15):2576–2585
46. Schmidt U, Abramova A, Boucheron N, Eckelhart E, Schebesta A, Bilic I et al (2009) The protein tyrosine kinase Tec regulates mast cell function. Eur J Immunol 39(11):3228–3238
47. Iyer AS, Morales JL, Huang W, Ojo F, Ning G, Wills E et al (2011) Absence of Tec family kinases interleukin-2 inducible T cell kinase (Itk) and Bruton's tyrosine kinase (Btk) severely impairs Fc epsilonRI-dependent mast cell responses. J Biol Chem 286(11):9503–9513
48. Iwaki S, Tkaczyk C, Satterthwaite AB, Halcomb K, Beaven MA, Metcalfe DD et al (2005) Btk plays a crucial role in the amplification of Fc epsilonRI-mediated mast cell activation by kit. J Biol Chem 280(48):40261–40270
49. Zorn CN, Keck S, Hendriks RW, Leitges M, Freudenberg MA, Huber M (2009) Bruton's tyrosine kinase is dispensable for the toll-like receptor-mediated activation of mast cells. Cell Signal 21(1):79–86
50. Luskova P, Draber P (2004) Modulation of the Fcepsilon receptor I signaling by tyrosine kinase inhibitors: search for therapeutic targets of inflammatory and allergy diseases. Curr Pharm Des 10(15):1727–1737
51. Gagliardi MC, Finocchi A, Orlandi P, Cursi L, Cancrini C, Moschese V et al (2003) Bruton's tyrosine kinase defect in dendritic cells from X-linked agammaglobulinaemia patients does not influence their differentiation, maturation and antigen-presenting cell function. Clin Exp Immunol 133(1):115–122
52. Liu Y, Wu Y, Lam KT, Lee PP, Tu W, Lau YL (2012) Dendritic and T cell response to influenza is normal in the patients with X-linked agammaglobulinemia. J Clin Immunol 32(3):421–429
53. Kawakami Y, Inagaki N, Salek-Ardakani S, Kitaura J, Tanaka H, Nagao K et al (2006) Regulation of dendritic cell maturation and function by Bruton's tyrosine kinase via IL-10 and Stat3. Proc Natl Acad Sci U S A 103(1):153–158
54. Singhal E, Kumar P, Sen P (2011) A novel role for Bruton's tyrosine kinase in hepatocyte growth factor-mediated immunoregulation of dendritic cells. J Biol Chem 286(37):32054–32063
55. Kurz SM, Diebold SS, Hieronymus T, Gust TC, Bartunek P, Sachs M et al (2002) The impact of c-met/scatter factor receptor on dendritic cell migration. Eur J Immunol 32(7):1832–1838
56. Lougaris V, Baronio M, Vitali M, Tampella G, Cattalini M, Tassone L et al (2014) Bruton tyrosine kinase mediates TLR9-dependent human dendritic cell activation. J Allergy Clin Immunol 133(6):1644–50.e4
57. Rock J, Schneider E, Grun JR, Grutzkau A, Kuppers R, Schmitz J et al (2007) CD303 (BDCA-2) signals in plasmacytoid dendritic cells via a BCR-like signalosome involving Syk, Slp65 and PLCgamma2. Eur J Immunol 37(12):3564–3575
58. Nakayama M, Akiba H, Takeda K, Kojima Y, Hashiguchi M, Azuma M et al (2009) Tim-3 mediates phagocytosis of apoptotic cells and cross-presentation. Blood 113(16):3821–3830
59. Maurya N, Gujar R, Gupta M, Yadav V, Verma S, Sen P (2014) Immunoregulation of dendritic cells by the receptor T cell Ig and mucin protein-3 via Bruton's tyrosine kinase and c-Src. J Immunol 193(7):3417–3425
60. Jefferies CA, Doyle S, Brunner C, Dunne A, Brint E, Wietek C et al (2003) Bruton's tyrosine kinase is a toll/interleukin-1 receptor domain-binding protein that participates in nuclear factor kappaB activation by Toll-like receptor 4. J Biol Chem 278(28):26258–26264
61. Lopez-Herrera G, Vargas-Hernandez A, Gonzalez-Serrano ME, Berron-Ruiz L, Rodriguez-Alba JC, Espinosa-Rosales F et al (2014) Bruton's tyrosine kinase--an integral protein of B cell development that also has an essential role in the innate immune system. J Leukoc Biol 95(2):243–250
62. Mansell A, Smith R, Doyle SL, Gray P, Fenner JE, Crack PJ et al (2006) Suppressor of cytokine signaling 1 negatively regulates toll-like receptor signaling by mediating Mal degradation. Nat Immunol 7(2):148–155
63. Fiedler K, Sindrilaru A, Terszowski G, Kokai E, Feyerabend TB, Bullinger L et al (2011) Neutrophil development and function critically depend on Bruton tyrosine kinase in a mouse model of X-linked agammaglobulinemia. Blood 117(4):1329–1339
64. Marron TU, Rohr K, Martinez-Gallo M, Yu J, Cunningham-Rundles C (2010) TLR signaling and effector functions are intact in XLA neutrophils. Clin Immunol 137(1):74–80

65. Honda F, Kano H, Kanegane H, Nonoyama S, Kim ES, Lee SK et al (2012) The kinase Btk negatively regulates the production of reactive oxygen species and stimulation-induced apoptosis in human neutrophils. Nat Immunol 13(4):369–378
66. Krupa A, Fudala R, Florence JM, Tucker T, Allen TC, Standiford TJ et al (2013) Bruton's tyrosine kinase mediates FcgammaRIIa/toll-like receptor-4 receptor crosstalk in human neutrophils. Am J Respir Cell Mol Biol 48(2):240–249
67. Hartkamp LM, Fine JS, van Es IE, Tang MW, Smith M, Woods J et al (2014) Btk inhibition suppresses agonist-induced human macrophage activation and inflammatory gene expression in RA synovial tissue explants. Ann Rheum Dis
68. Whang JA, Chang BY (2014) Bruton's tyrosine kinase inhibitors for the treatment of rheumatoid arthritis. Drug Discov Today 19(8):1200–1204
69. Mueller H, Stadtmann A, Van Aken H, Hirsch E, Wang D, Ley K et al (2010) Tyrosine kinase Btk regulates E-selectin-mediated integrin activation and neutrophil recruitment by controlling phospholipase C (PLC) gamma2 and PI3Kgamma pathways. Blood 115(15):3118–3127
70. Gilbert C, Levasseur S, Desaulniers P, Dusseault AA, Thibault N, Bourgoin SG et al (2003) Chemotactic factor-induced recruitment and activation of Tec family kinases in human neutrophils. II. Effects of LFM-A13, a specific Btk inhibitor. J Immunol 170(10):5235–5243
71. Lachance G, Levasseur S, Naccache PH (2002) Chemotactic factor-induced recruitment and activation of Tec family kinases in human neutrophils. Implication of phosphatidynositol 3-kinases. J Biol Chem 277(24):21537–21541
72. Block H, Zarbock A (2012) The role of the tec kinase Bruton's tyrosine kinase (Btk) in leukocyte recruitment. Int Rev Immunol 31(2):104–118
73. Krupa A, Fol M, Rahman M, Stokes KY, Florence JM, Leskov IL et al (2014) Silencing Bruton's tyrosine kinase in alveolar neutrophils protects mice from LPS/immune complex-induced acute lung injury. Am J Physiol Lung Cell Mol Physiol 307(6):L435–L448
74. Horwood NJ, Mahon T, McDaid JP, Campbell J, Mano H, Brennan FM et al (2003) Bruton's tyrosine kinase is required for lipopolysaccharide-induced tumor necrosis factor alpha production. J Exp Med 197(12):1603–1611
75. Gray P, Dunne A, Brikos C, Jefferies CA, Doyle SL, O'Neill LA (2006) MyD88 adapter-like (Mal) is phosphorylated by Bruton's tyrosine kinase during TLR2 and TLR4 signal transduction. J Biol Chem 281(15):10489–10495
76. Mangla A, Khare A, Vineeth V, Panday NN, Mukhopadhyay A, Ravindran B et al (2004) Pleiotropic consequences of Bruton tyrosine kinase deficiency in myeloid lineages lead to poor inflammatory responses. Blood 104(4):1191–1197
77. Mukhopadhyay S, Mohanty M, Mangla A, George A, Bal V, Rath S et al (2002) Macrophage effector functions controlled by Bruton's tyrosine kinase are more crucial than the cytokine balance of T cell responses for microfilarial clearance. J Immunol 168(6):2914–2921
78. Perez de Diego R, Lopez-Granados E, Pozo M, Rodriguez C, Sabina P, Ferreira A et al (2006) Bruton's tyrosine kinase is not essential for LPS-induced activation of human monocytes. J Allergy Clin Immunol 117(6):1462–1469
79. Jansson L, Holmdahl R (1993) Genes on the X chromosome affect development of collagen-induced arthritis in mice. Clin Exp Immunol 94(3):459–465
80. Schmidt NW, Thieu VT, Mann BA, Ahyi AN, Kaplan MH (2006) Bruton's tyrosine kinase is required for TLR-induced IL-10 production. J Immunol 177(10):7203–7210
81. Cheng G, Ye ZS, Baltimore D (1994) Binding of Bruton's tyrosine kinase to Fyn, Lyn, or Hck through a Src homology 3 domain-mediated interaction. Proc Natl Acad Sci U S A 91(17):8152–8155
82. Lee HS, Moon C, Lee HW, Park EM, Cho MS, Kang JL (2007) Src tyrosine kinases mediate activations of NF-kappaB and integrin signal during lipopolysaccharide-induced acute lung injury. J Immunol 179(10):7001–7011
83. Jefferies CA, O'Neill LA (2004) Bruton's tyrosine kinase (Btk)-the critical tyrosine kinase in LPS signalling? Immunol Lett 92(1-2):15–22

84. Lee KG, Xu S, Kang ZH, Huo J, Huang M, Liu D et al (2012) Bruton's tyrosine kinase phosphorylates toll-like receptor 3 to initiate antiviral response. Proc Natl Acad Sci U S A 109(15):5791–5796
85. Ni Gabhann J, Hams E, Smith S, Wynne C, Byrne JC, Brennan K et al (2014) Btk regulates macrophage polarization in response to lipopolysaccharide. PLoS One 9(1), e85834
86. Koprulu AD, Kastner R, Wienerroither S, Lassnig C, Putz EM, Majer O et al (2013) The tyrosine kinase Btk regulates the macrophage response to Listeria monocytogenes infection. PLoS One 8(3), e60476
87. Gonzalez-Serrano ME, Estrada-Garcia I, Mogica-Martinez D, Gonzalez-Garay A, Lopez-Herrera G, Berron-Ruiz L et al (2012) Increased pro-inflammatory cytokine production after lipopolysaccharide stimulation in patients with X-linked agammaglobulinemia. J Clin Immunol 32(5):967–974
88. Khare A, Viswanathan B, Gund R, Jain N, Ravindran B, George A et al (2011) Role of Bruton's tyrosine kinase in macrophage apoptosis. Apoptosis Int J Progr Cell Death 16(4):334–346
89. Byrne JC, Ni Gabhann J, Stacey KB, Coffey BM, McCarthy E, Thomas W et al (2013) Bruton's tyrosine kinase is required for apoptotic cell uptake via regulating the phosphorylation and localization of calreticulin. J Immunol 190(10):5207–5215
90. Fluckiger AC, Li Z, Kato RM, Wahl MI, Ochs HD, Longnecker R et al (1998) Btk/Tec kinases regulate sustained increases in intracellular Ca2+ following B-cell receptor activation. EMBO J 17(7):1973–1985

Index

A. Plebani, V. Lougaris (eds.), *Agammaglobulinemia*,
Rare Diseases of the Immune System 4, DOI 10.1007/978-3-319-22714-6

MIX
Papier aus verantwortungsvollen Quellen
Paper from responsible sources
FSC® C105338

If you have any concerns about our products,
you can contact us on
ProductSafety@springernature.com

In case Publisher is established outside the EU,
the EU authorized representative is:
Springer Nature Customer Service Center GmbH
Europaplatz 3, 69115 Heidelberg, Germany

Printed by Libri Plureos GmbH
in Hamburg, Germany